TESTABILITY CONCEPTS FOR DIGITAL ICs

Frontiers in Electronic Testing

Volume 3

Testability Concepts for Digital ICs

The Macro Test Approach

by

F. P. M. Beenker
Philips Medical Systems
(formerly Philips Research)

R. G. Bennetts
Synopsys, Inc.

and

A. P. Thijssen
TU Delft

SPRINGER SCIENCE+BUSINESS MEDIA, B.V.

A C.I.P. Catalogue record for this book is available from the Library of Congress

ISBN 978-1-4613-6004-9 ISBN 978-1-4615-2365-9 (eBook)
DOI 10.1007/978-1-4615-2365-9

Printed on acid-free paper

Table of Contents

Preface vii

1 Introduction **1**
- 1.1 The Main Topic 1
- 1.2 Test Objectives 2
- 1.3 Definition of Testability 6
- 1.4 Problem Statement: Strategies and Requirements 7
- 1.5 Outline 7

2 Defect-Oriented Testing **9**
- 2.1 Reason 9
- 2.2 Defects and Faults 11
- 2.3 Defect-Fault Relationship: Inductive Fault Analysis 12
- 2.4 Fault-Defect Relationship: Process Monitoring Testing 14

3 Macro Test: A Framework for Testable IC Design **19**
- 3.1 Introduction to the Macro Test Philosophy 19
  - 3.1.1 Macro Test driven by Quality Requirements 19
  - 3.1.2 Macro Test driven by IC Design Styles 21
  - 3.1.3 The Macro Test Concepts 23
  - 3.1.4 Macro Definition 30
- 3.2 Testability Synthesis within the Macro Test Concept 31
- 3.3 Integration of Macro Test into a Design & Test flow 35
  - 3.3.1 The Evaluation Plan 35
  - 3.3.2 Interfacing & Integrating 37
- 3.4 Summary of Essential Macro Test Items 39

4 Examples of Leaf-Macro Test Techniques **41**
- 4.1 Defect Modeling and Test Algorithm Development for Static Random Access Memories (SRAMs) 42
  - 4.1.1 Introduction 42
  - 4.1.2 Development of an SRAM Fault Model 44
  - 4.1.3 Fault Propagation 52
  - 4.1.4 The SRAM Test Algorithm 54
  - 4.1.5 Practical Validation 58
  - 4.1.6 Conclusions 62
- 4.2 Built-In Self-Test for Static Random Access Memories 65
  - 4.2.1 Introduction and Motivation 65
  - 4.2.2 Specification and Architecture of the Self-Test Machine 65
  - 4.2.3 The Various Blocks of the Self-Test Machine 69
- 4.3 Leaf-Macro Testability Study Aspects 76

5 Scan Chain Routing with Minimal Test Application Time 81
5.1 Leaf-Macro Access 81
5.2 Introduction to Scan Chain Routing 83
5.3 Scan Test Application Protocol 85
5.4 Scan Chain Routing Problem Formulation 86
5.5 Scan Chain Routing Cost Model 88
5.6 Scan Chain Routing Problem Complexity 92
5.7 Routing of Scan Registers into a Single Scan Chain 96

6 Test Control Block Concepts 107
6.1 Introduction 107
6.2 Test Control Block Requirements 108
6.3 Test Controller Architectures 108
6.4 Relation between a Test Control Block and Test Plans 114
6.5 Test Control Block Design Requirements 117
6.6 Optimal Test Control Block implementation 120
6.6.1 TCB Optimization via State Merging 123
6.6.2 TCB Optimization via State Assignment 124
6.6.3 TCB Optimization via Specification of Unused States 128
6.7 Test Control Block Design Example 129
6.8 Distributed Test Control 134

7 Exploiting Parallelism in Leaf-Macro Access 139
7.1 Introduction 139
7.2 Levels of Parallelism 140
7.3 Formal Definitions of Resources, Resource Compatibility and Parallelism 150
7.4 Test Compatibility Graphs 162
7.5 Resource Allocation versus Test Assembly 163
7.6 Algorithmic Implementation and Experimental Results 166

8 Timing Aspects of CMOS VLSI Circuits 171
8.1 Introduction 171
8.2 Timing Models of Latches and Flip-Flops 176
8.3 Timing of Data Transfers 180
8.4 Clock Drivers 189

List of Symbols and Abbreviations 193

References 197

Index 207

Preface

Testing Integrated Circuits for manufacturing defects includes four basic disciplines. First of all an understanding of the origin and behaviour of defects. Secondly, knowledge of IC design and IC design styles. Thirdly, knowledge of how to create a test program for an IC which is targeted on detecting these defects, and finally, understanding of the hardware, Automatic Test Equipment, to run the test on. All four items have to be treated, managed, and to a great extent integrated before the term 'IC quality' gets a certain meaning and a test a certain measurable value.

The contents of this book reflects our activities on testability concepts for complex digital ICs as performed at Philips Research Laboratories in Eindhoven, The Netherlands. Based on the statements above, we have worked along a long-term plan, which was based on four pillars.

1. The definition of a test methodology suitable for 'future' IC design styles,
2. capable of handling improved defect models,
3. supported by software tools, and
4. providing an easy link to Automatic Test Equipment.

The reasoning we have followed was continuously focused on IC quality. Quality expressed in terms of the ability of delivering a customer a device with no residual manufacturing defects. Bad devices should not escape a test.

The basis of IC quality is a thorough understanding of defects and defect models. Research was therefore started on defect modeling and after some period of time help was obtained from Carnegie Mellon University in Pittsburgh. A variety of design modules were studied ranging from PLAs, memories, standard cells, and a multiplier. The detailed ground work was mostly done by students from the University of Eindhoven and the University of Delft. Defect models were developed, specific tests were generated and design requirements for specific test solutions were stated. This study resulted in the awareness that testing plays an important role in the quality improvement process for design and manufacturing. Testing has the role to measure and to provide corrective feedback to improve design and manufacturing processes. The choice for the application of the Inductive Fault Analysis technique has been correct. We did not study theoretical fault models but focused our effort on realistic defect modeling methods. One of the major results of this period was the development of memory defect models, algorithms and self-test techniques. The resulting SRAM memory test algorithm is currently in wide-spread use. The continuing research on defect modeling has also geared the research on I_{DDQ}, design centering, and technology centering capabilities.

Having obtained a detailed understanding of how to test a specific design module, henceforth called a macro, the next problem to solve was how to test a macro embedded in the device. A period of brainstorming and specification resulted in the details of the Macro Test concepts for test data access and test control access. The basic ideas were formulated and an initial set of Macro Test supporting software tools was implemented. This implementation, known as Sphinx, showed the feasibility of the Macro Test concepts; [Beenker89, Claasen89, Beenker90, Woudsma90]. Via the Esprit Everest project, the Jessi Sigma project, dedicated marketing effort, and internal Philips projects, the Sphinx prototype has emerged into software products currently in use.

Acknowledgements

We had the good fortune to work with many interesting people during our period at Philips Research working on the Macro Test ideas.

A number of colleagues have produced excellent work. The work on SRAM defect modeling was started by Pieter Veenstra. Based on his results, we concluded that a more systematic approach towards defect modeling was required and we decided to apply the Inductive Fault Analysis technique. The details of the application of this technique on an SRAM and the implementation of the resulting algorithm was Rob Dekker's graduating study. Michiel Ligthart performed the analysis on PLAs. Marcel Tjin installed the IFA software of CMU at the Philips Research Laboratories and applied the software for an Inductive Fault Analysis of some standard cells. Erik-Jan Marinissen studied optimization techniques for Test Control Block designs and Frank Bouwman improved Erik-Jan's results. Steven Oostdijk formulated the initial theory of scan chain routing and Hans Bouwmeester started the theoretical analysis of test time optimization. We have put all this work into perspective and detailed the theoretical aspects.

We would like to thank Eric van Utteren, Jan Janse, and Theo Claasen who gave the opportunity to work on this topic for such a long time.

We worked along a plan which was written in 1985 and which for a major portion is still valid. This plan could not have been written and implemented without the support of Karel van Eerdewijk, Frank Peacock, and Rudi Stans.

The development of the Macro Test software has been a great experience. A team of highly motivated people contributed to this success. Besides initially Tim Murphy, Daniel Vangheluwe and Rudi Stans, this team consisted of Frank Bouwman, Steven Oostdijk, Frank van Latum, Marc van de Velden, Taco Brinkhoff, and John Zijlstra.

Much of the ground work on defect-oriented testing was performed by Rob Dekker, Erik Bruls, Fred Camerik and Frank Agricola with the great help of Carnegie Mellon University (Prof. Wojciech Maly and his student Samir Naik), Technical University of Eindhoven (Prof. Jess' group), University of Barcelona (Joan Figueras' group) and INESC (Paolo Teixeira's group).

We appreciated the many lively discussions on timing aspects with Bas Samsom of the Delft University of Technology.

We would like to thank our colleagues and especially Keith Baker for the many highly interesting discussions we have had; Emile Aarts for his help on the mathematical aspects of the Macro test theory; our colleagues Engel Roza's group for their cooperative spirit during the design projects; and Max van der Star who contributed a lot to the development of the conceptual Macro Test ideas.

During this period, we have been in contact with a wide variety of international companies, universities and individual people. We have learned a lot from the discussions with all these professional people.

Frans P.M. Beenker
'Ben' R.G. Bennetts
Loek A.P. Thijssen

1 Introduction

1.1 The Main Topic

Throughout the 1980s and 1990s, the theory and practice of testing electronic products have changed considerably. As a consequence of exploiting the ever more advanced technologies, the complexities of products have increased significantly and so have the testing problems. Testing has become fundamental to the design, manufacture and delivery of a quality product.

The generation of high quality tests, with respect to all kinds of requirements, has become complex and time consuming. The problem is becoming even more complex now that modern IC design tools are causing the variety of products to increase rapidly. It is unacceptable that the time required for test development is an order of magnitude more than the time needed for design. Therefore, we have to consider the role and the responsibilities of testing across the entire organization and product development process in order to achieve significant reduction in time and costs.

Requirements such as fast design, high product quality, and reliability, reflect the demands imposed on test strategies. However, most organizations do not focus on one single product. Usually, a whole range of products over a wide variety of product classes is continuously being developed. Each product has its own specific test problems and test departments quickly become overloaded with a continuous flow of different products to be tested and evaluated.

Technologies keep changing and soon today's methods of testing and evaluating products will no longer meet the requirements. This situation is not ideal insofar as testing is concerned. And it certainly is not in the case where testing is considered to be a side activity; the 'throw-it-over-the-wall' attitude. The only way to overcome this problem is to change the environment and to reorganize the design and test activities in such a way that testing can keep up with the design activities, i.e. the integrated solution. This is more feasible in organizations that are able to keep a tight control of all the stages of design and manufacturing. In such cases, a coherent framework for testing can be developed all along the design trajectory [Claasen89].

A manufacturer of quality products supplies a 'known good product' that performs to a given specification. Here we reach the bottom line. Quality and testing are indissolubly linked and are both fundamental to the generation of revenue to a company, helping the company to remain profitable and therefore to survive. Testing plays an important role in assessing the quality of a product. The tester acts as a filter, filtering good products from bad. Unfortunately, the tester can pass bad products and fail good products.

We can quantify quality by the twin requirements of *zero escapes* (no bad product passes a test) and *zero defects* (the product is manufactured without a defect). The former requires the test environment to possess the property of 'excellent defect detection', whereas the latter requires the property of 'excellent defect location'. A zero defects target can only be achieved when defects that occur are analyzed and prevented from occurring again. This means that the product yield goes up. Zero defects and zero escapes, combined with fast design and a quickly climbing yield curve, are necessities for a company to develop a good market position. These dual targets impose such requirements as fast and accurate test pattern generation, automatic test program generation, and minimal testing costs. The requirements are usually conflicting and an optimal solution has to be found within a search space that cannot easily be explored. The discipline capable of achieving this solution is called Design-for-Test (DfT). The basic idea behind DfT is that during the design phase the test requirements are already been taken into account. Thus the design and test activities become integrated, thereby effecting a manageable test program route. Although such a route is manageable, this does not necessarily mean that the final outcome meets all requirements. Commonly used DfT methods such as scan design [Eichelbrg78, Eichelbrg91] and partial scan [Trischler80] do not fully explore the search space. *A DfT method, which provides a manageable test program route, and which is capable of exploring the search space is the main topic of this book.*

1.2 Test Objectives

We can formulate the following global *test objectives.*

* To check that the product works according to its specifications.
* If it does not work, locate the cause of failure.
* Feed back accurate information to eliminate the cause of the fault in the future.

The first test objective has to be mapped onto the various product development stages and product quality requirements. The product development stages range from requirement specification to functional specification, functional design, and structural design to implementation. These stages are indicated on the left part of Figure 1.1.

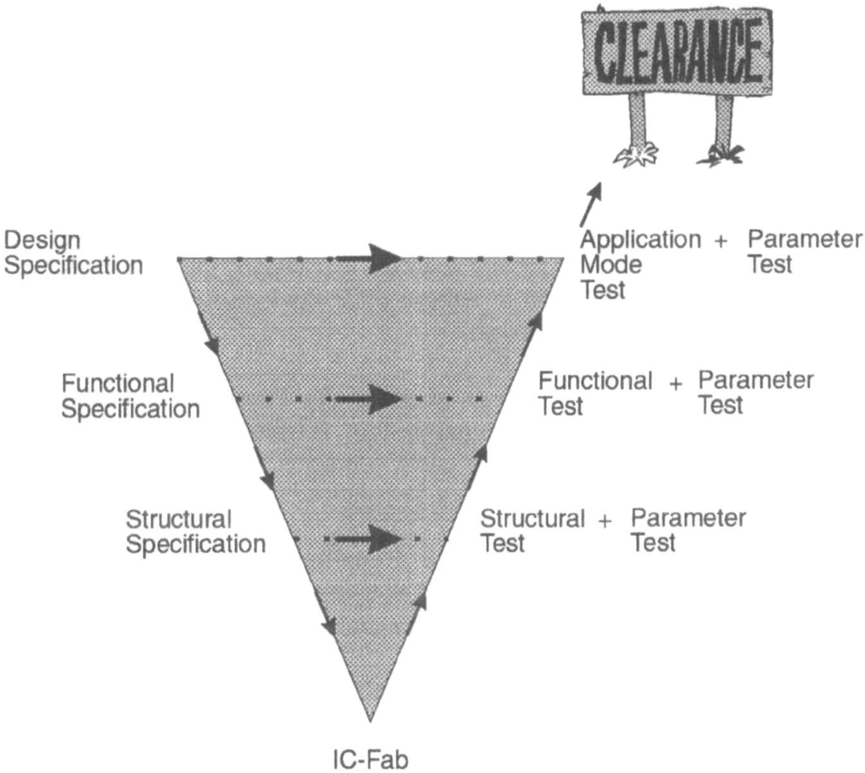

Figure 1.1. An idealized verification process.

The most uncertain stages are those from requirement specification to functional specification, and the IC implementation. These stages always have to be verified. All other stages can either be proved to be correct or are verified during the design by means of simulation tools and design rule checkers. Iterations during design are made and finally the product is manufactured. This is followed by extensive *verification testing*. Verification testing usually takes place in several stages: each stage has a separate purpose, and increases the confidence that the IC has been implemented to specification. The concept is illustrated in Figure 1.1. Successive levels of testing are indicated that check each level from the specification to the implementation. Different methods are used at different levels because of the nature of the specification.

To summarize, the successive levels of testing are

1. *structural testing*,
2. *functional testing*,
3. *application-mode testing*, and
4. *parameter testing*.

At the end of the verification process, with as few time-consuming design and manufacturing iterations as possible, the design is ready and the final production test is run. This test is required to remove those devices that have been malformed during the fabrication process.

The test-level categories are often named differently. Functional testing may be called functional validation. Structural testing is sometimes called process verification or fault-effect testing. To avoid confusion, we use the terms listed above.

From Figure 1.1 it can be deduced that a number of people are necessarily involved in the design and test process. The systems designer and IC designer focus on the design part and the test engineer focuses on the test part.

These people all have a different view on testing. The system designers are concerned with the application correctness of the designs. The IC designers are concerned about the one-to-one relationship between their implementation and the functional specification. The test engineer is chiefly concerned about the correct functioning of the circuit according to the specifications and the correct manufacturing of the product according to the relevant defect spectrum.

Structural Testing

Structural testing addresses the question of whether the product was built according to the original structural specification and whether any defects were introduced by the manufacturing process itself.

Structural testing requires a detailed knowledge of the defects which occur during the IC processing stages. Those defects which cause abnormal functional behavior, must be detected during the structural test. Testing for all possible faults is not feasible because of the limitations of test execution time. Hence, trade-off decisions based on an analysis of the risk of passing a faulty device or board have to be made. The derived set of defect behaviors constitutes a *fault model. Test patterns* have to be generated, applied to the device and evaluated for all faults in the fault model. It will be clear that the choice of a fault model is most critical for the quality of a test. Understanding the relationship between IC defects and their corresponding fault models is difficult. Techniques to improve fault models are available and a general introduction is given in Chapter 2.

Functional testing

Functional testing concentrates on the question of the functionality requirements of the product: 'Does the product do what it is supposed to do?' What this usually means is testing the product at the critical ends of its specifications. Test data for functional testing is not created by analysis of the IC's structure, but by the designer who produces verification tests to prove the design's compliance with a higher-level specification. The design verification data base is a rich source of test data that provides the functional testing process with essential information. Functional testing demands interactive testing. In order to verify the product, the test engineer is able to vary test patterns interactively and to observe the results.

Application-mode testing

For complex ICs, neither of the above methods are alone sufficient and more exhaustive methods are required. Simulation results in general cannot provide sufficient test data to simulate a design in all modes of application. Another reason to require more exhaustive methods is that each test method checks its own specific level. The step from requirement specification to functional specification is never really checked. The application-mode testing covers this step. During application-mode testing the ICs are tested with real-life data in a wide variety of the environmental conditions it is expected to encounter during its normal life cycle.

Application-mode testing could be done by creating a test rig to recreate the applications environment. However, creating test rigs is a time-consuming business. Moreover, a test rig is a difficult environment in which to debug design problems. The solution is a 'virtual' application-mode testing, where software models are used to create and animate visualization of the applications mode. The results can then be run on a class of IC testers called verification testers (HP82000, IMS-ATS, etc.) [Mehtani92].

Parameter testing

One of the requirements for ensuring parameter correctness is that the ICs are working under various parameter conditions and environmental constraints. The parameter test puts emphasis on issues such as the frequency of operation, acceptable tolerances on power supply, temperature ranges, and power dissipation. The parameter tests are used in combination with structural, functional or application-mode test data.

In this book we pay attention only to structural testing objectives.

1.3 Definition of Testability

Having defined the test requirements and the test objectives, we are able to define the term Testability of Digital ICs [Bennetts84].

A digital IC is testable if test patterns can be generated, applied, and evaluated in such a way as to satisfy pre-defined levels of performance (e.g. detection, location, application) within a pre-defined cost budget and time scale.

This definition is still vague and open to subjective interpretation. Each company has different quality requirements, different cost structures, different products and different time scales. What is needed for a proper interpretation of the definition is a thorough understanding of the capabilities and limitations of the various methods, tools and techniques required to produce a working test program. This also means that a generic solution which will satisfy all the requirements of every company cannot exist. Therefore, methods are required which provide the possibility to fine-tune and optimize for use in a certain application, product and environment. *Macro Test* is such a method and it is the main theme of this book.

As with every other Design-for-Testability method, Macro Test requires that there always well defined timing intervals where signals can be applied and observed. This in itself requires a synchronous timing strategy, at least during test for the IC part to be tested. Attention to this special, and often ignored, topic is given in Chapter 8.

1.4 Problem Statement: Strategies and Requirements

We have defined the term '*testability*'. This definition embraces a number of issues such as the aspects of quality, economics and characterization of the software, hardware and 'human'ware. The implementation of testability in a large company is therefore not only a matter of buying or developing software tools. Matters which are of importance in the implementation of testability have to do with

* organization and responsibilities,
* the visibility of the test problems,
* the development of strategies and methodologies to solve the test problems and meet the targets of quality and cost,
* insight into the available IC design methodology in order to be able to optimize for a certain product or product family,
* insight into the theory of testing,
* knowledge of limitations of the test hardware and software,
* knowledge of manufacturing processes and defect analysis, and, finally,
* a set of tools to support all the required activities and knowledge as to the effectiveness of the tools in relation to the application.

The strategy to be taken is to integrate the testability aspects into the design and manufacturing of ICs and to define for each IC design project precisely the boundary conditions, responsibilities, interfaces and communications between persons and quality targets. The Macro Test activities as such form an inherent part of the total set of activities and are primarily intended to maximize the quality and minimize the costs of the structural test activities.

1.5 Outline

The work as described in this book is a summary of the results of a 10-year period of research performed at the Philips Research Laboratory in Eindhoven, the Netherlands. The work has migrated towards a Design-for-Testability concept suitable for economical application on complex ICs. A wide variety of Macro Test aspects were considered during this period, most of which can be found in this book. For more details we refer to [Beenker94].

Issues that have to do with realistic defect modeling are described in Chapter 2. Chapter 3 provides an in-depth study on the thought behind Macro Test. Specific macro test methods and defect models are described in Chapter 4. One of the major areas of concern in Macro Test is accessibility from the device pins to embedded macros. Techniques to provide access from device pins to macros for test data and test control are the topic of Chapter 5 and Chapter 6, respectively. Optimization techniques to enable a minimal test application time are described in Chapter 7. Finally, timing strategies for reliable IC design, as a basic boundary condition for testable IC design, are the topic of Chapter 8.

2 Defect-Oriented Testing

2.1 Reason

The key problem in the electronics industry is the need to improve quality and productivity while reducing costs. This is a simple statement; however, it influences every step of a product-development cycle. For integrated circuits, these development cycles are the design, manufacture and testing cycles. All three cycles have their own specific demands on the information required to improve their productivity and yield. The information packages are clearly correlated; see Figure 2.1.

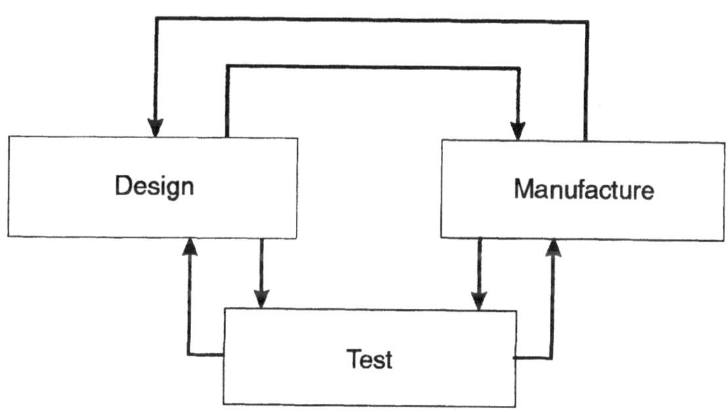

Figure 2.1. Correlation between design, manufacture and test.

A designer needs to know the technology in which a circuit is going to be manufactured and the layout style to use. On which criteria should the choice be made? If the designer were to base his criteria on minimal die size and hence, make the choice for the latest available technology, the upshot could very well be that the yield of the product would be too low to justify the effort. If the yield is too low then the product price rises, which could make the product economically non-viable. Therefore, there is a need to design a circuit in such a way as to maximize the yield and to minimize the costs which creates the need for yield estimation, the indication of yield sensitive places in the layout, etc. [Maly90].

In addition to production, in manufacturing, there is the task of developing processes which enable the manufacturing of components with an acceptable yield. When a new generation of technology is being developed, the learning curve climbing rate is relatively slow and the initial yield is low. The faster the manufacturing industry can climb the learning curve of a process, the higher will be the product yield, the lower its price, and the higher the potential market share. Hence, a company which is able to realize the fastest learning curve with a new process and new product will be the first to drop their prices for the product and will thus have the largest market share. To achieve a fast learning curve, information about the causes of yield drops is necessary.

Testing has become a central factor in gathering the information necessary for design and manufacture optimization. The application of tests to a product generates much data on both the product and the fabrication process. Based on the information gathered from the test of a device, the following two ways of reasoning are possible.

Design centering.
: Design the device in such a fashion that the device quality is insensitive to random process variations. For example, this can be done by adjusting device geometries and circuit topologies. In order to do this, information is required on the most yield-sensitive part of the device and an indication of the expected yield for this product in a given technology.

Technology centering.
: Improve the quality of the manufacturing process. This can be done by improving the quality of the raw material, the precision of the processing equipment and the cleanliness of the facilities. Information is required as to which part of the process, process parameters, or equipment, is the most critical yield-limiting factor. Can we measure the influence of parameter variations on the product itself and can we perform defect analysis on a large number of products in order to produce reliable statistical data?

Note here the potential value of test data and the corresponding central role of testing in improving the quality of a process and of a device. Exploiting this central role effectively is called *defect-oriented testing.*

This chapter provides a brief outline of the defect-oriented test techniques we have used during our research and which mainly came from Carnegie Mellon University [Shen85, Ferguson88a, Dekker88a]. We argue that the ability to apply defect-oriented testing is one of the key items of the Macro Test.

2.2 Defects and Faults

As explained in Chapter 1, we here focus our attention on structural testing. In this respect, a *defect* is considered to be a change in the geometry of a device. Defect-oriented testing requires information about the defects that occur. Various methods are used for process monitoring and for studying the cause of defects. Defect causes can be revealed by means of failure analysis techniques. Examples of such techniques are visual inspection and SEM inspection, on-chip probing, liquid crystal applications and de-processing. These techniques are indispensable but they are manually carried out and require extensive efforts. A few dies per day is about the maximum that can be handled. Automatic data-gathering systems are applied to speed up the process of analyzing defects and gathering defect data and to obtain statistical data based on many measurements. Examples of such automatic systems are electrical defect monitors [Bruls91] and automatic visual inspection systems or a combination of both.

The defect analysis activities focus their attention on the following two categories of defects at layout level.

* *Global defects*. For instance, too-thick gate oxide or too-thin poly silicon caused by process etching errors or oven temperature variations, the misalignment of masks, variants in dopant distributions, and deviations from the designed dimensions.
* *Spot defects*. For instance, dust particles on the chip or the masks, scratches and gate oxide pinholes.

The impact of global defects is extensive. Hence, they are detected before structural testing by using well-defined and widely used *Process Control Modules* (PCMs) [Swaving88]. A vast majority of defects are caused by local spot defects [Maly85, Syrzycki87]. For this reason, we consider spot defects only for our defect-oriented testing purposes.

Spot defects are studied by using monitor wafers. These wafers contain only defect monitors. There is a wide range of defect monitors, each providing some specific information about the defects. An example of a generic type of defect monitor in use is the *VLSI MIX* (VMX) monitor [Bruls91, Bruls93]. It contains essentially of three conducting layers with all three layers built up of a comb-string-comb structure. The VMX monitor can be used for both process- and product-related defect studies. Typical information from these monitor wafers includes *Defect Size Distribution* and *Defect Density Distribution.* A Defect Size Distribution defines the probability of occurrence of a defect of a certain size. A Defect Density Distribution defines the probability of occurrence of a defect in a certain layer of the IC structure.

For computational purposes, the spot defects are modeled as two-dimensional circles or as three-dimensional spheres [Pineda91]. For our purposes, we make use of a two-dimensional representation, which allows a defect study per IC layer. Depending on the processing step, spot defects be either missing or extra material on the IC structure. Missing and extra material may cause broken wires (opens), missing connections between two conducting layers (missing contact holes), connected wires (shorts), missing or extra transistors, etc.

We differentiate between a *defect* and a *fault.* Testing involves applying data to a device and measuring the electrical behaviour of the device. A change in the geometry of a device may result in a change in its functioning. The first, a change in geometry, is called a defect. The latter, a change in functionality, is called a fault. It will be clear that many defects can produce the same fault. It is also possible that a defect does not produce a fault.

We distinguish between *catastrophic* and *non-catastrophic defects.* Catastrophic defects cause a real break and a real short. Non-catastrophic defects cause a nearly broken connection and nearly a short. In our analysis we consider only catastrophic defects.

2.3 Defect-Fault Relationship: Inductive Fault Analysis

A structural test of a device must detect manufacturing defects. The normal practice in test program generation is that test engineers make use of fault models when there is expectation that the fault model covers a great number of the manufacturing defects. For defect-oriented testing, however, it is necessary that there is a much better understanding of the relationship between defect and faults and that the structural test does indeed cover the important manufacturing defects. This understanding is obtained with a methodology called *Inductive Fault Analysis* [Shen85, Ferguson88a, Ferguson88b]. The basics of the method are simple: the modeling of manufacturing defects in the layout of a device and extracting the behavior of the device in the presence of the defect. Inductive Fault Analysis is essentially a translation process from defects in a technology description to faults in a functional description.

The steps in the Inductive Fault Analysis approach are as follows.

1 *Computation of the physical characteristics of defects*
The physical characteristics of a defect, i.e., its size, location, and the layer of the IC structure in which the defect occurs are determined. The size and location of a defect are statistical data derived from a manufacturing line. The size is defined by the Defect Size Distribution (DSD) and the location is defined by the Defect Density Distribution (DDD). The Defect Size Distribution defines the probability of the occurrence of a spot defect of a certain size (e.g., the size is defined by the radius of the circle modeling the defect). The Defect Density Distribution defines the probability of the occurrence of a defect in a certain layer of the IC structure. It further defines how the probability of a defect depends on its size. DDDs and DSDs are available for each layer of an IC structure and derived via VMX and PCM analysis.

Considering all possible local changes in the geometry of the device in all layers will reveal a large number of computed defects.

2. *Placement of the defects on the layout*
As explained, we deal only with catastrophic defects. Step 2 determines whether the computed defect on a layer of the IC structure is a catastrophic defect and whether it is capable of altering the electrical behavior of the device.

The following two ways of dealing with defect computation and defect placement can be distinguished.

Randomized placement.
Using the DSDs and DDDs, defects can be computed and placed randomly in the layout. Each time a defect is generated which has the size and probability described in the DSDs and DDDs. Per defect, it is analyzed as to whether it will result in a fault.
Deterministic Placement.
The *'critical areas'*, the layout areas where a defect of a certain size will cause a catastrophic defect, can be calculated analytically [Pineda89, Pineda91]. The critical areas for all layers of material and for all sizes of defects can be related to the DDDs and DSDs. Hence, the computation of defects and placement in the layout can be done systematically.

3. *Extraction of the electrical equivalent of the defective circuit.*
This step identifies the electrical properties of the circuit. The result is a fault, i.e., the electrical equivalent of a defect. Note that many defects may be an element of the same fault equivalence class. Thus, a relationship between defects and faults is established. The list of faults derived from this procedure is weighted. For each fault its associated frequency of occurrence, or the probability of occurrence of the defects arizing this fault, in the physical circuit is estimated. Faults are modeled with extra transistors in the extracted transistor network. Thus, the effect of the fault can be switched on and off in a simulator by controlling the gate voltages of the extra transistors.

The test data of the defective circuit can be simulated employing a switch level simulator or a detailed transistor level simulator. Thus, a defect-fault simulator can be set up to analyze the defect detection capabilities of a test program for a certain device [Gracio89]. The functional behavior of a fault can be measured in terms of voltage and current. In this respect, defect-oriented testing has formed the basis for the re-newed attention to I_{DDQ} monitoring [Maly88, Soden89, Rodriguez91].

The Inductive Fault Analysis approach has been used to determine an effective test algorithm for Static Random Access Memories [Dekker88a]; see Chapter 4. One of the major results of this study was the development of a suitable fault model for SRAMs. It has been proven that only 50% of the possible defects could be detected when using a test algorithm based on the single stuck-at fault model.

2.4 Fault-Defect Relationship: Process Monitoring Testing

A defect-fault relationship makes it possible to generate a test program with a high defect detection capability. A fault-defect relationship is required for diagnosing the cause of failure of a device [Maly89]. The ideas are again easy to explain. Following the process of Inductive Fault Analysis allows a fault-defect relationship which allowed the development of a test program. By following the reverse arguments, we should be able to deduce from the obtained test data a list of the defects that cause the non-correct behavior. This twin principle is explained in Figure 2.2.

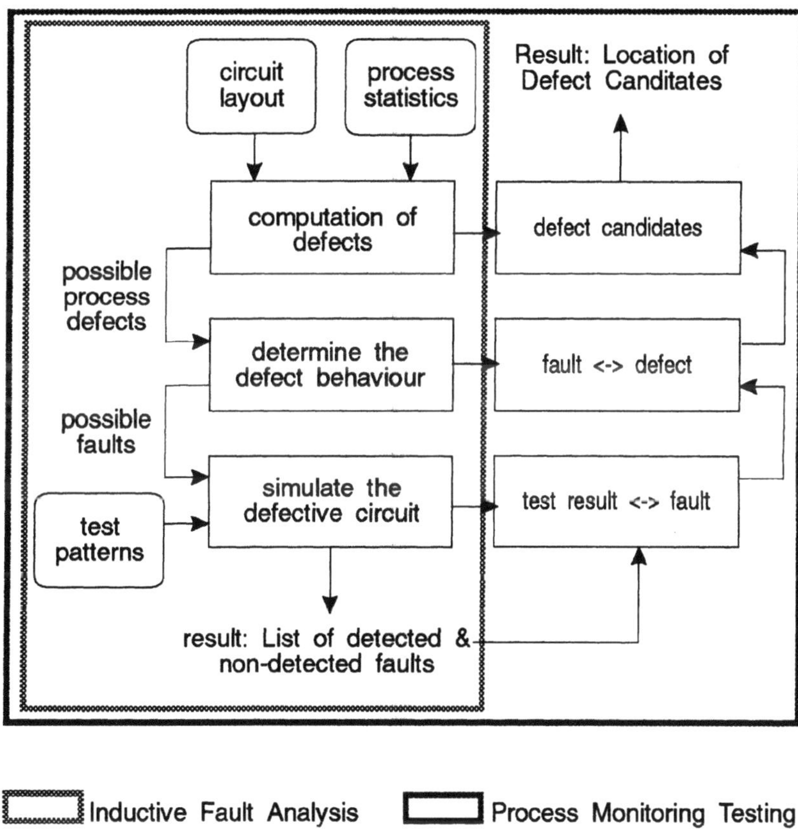

Figure 2.2. Inductive Fault Analysis and Process Monitoring Testing.

From Figure 2.2 it can be seen that the concept of Process Monitoring Testing is an extension of the Inductive Fault Analysis procedure. The main purpose is to obtain insofar as possible a unique test result for a certain defect, which provides the best resolution. The following sequence of steps has to be performed.

1. *Computation of defects.*
2. *Placement of defects in the layout.*
3. *Extraction of the electrical diagram of the defective circuit.*

   So far the procedure is identical to the Inductive Fault Analysis procedure. To enhance diagnostic's capabilities the following steps are added.

4. *Simulation of the defective IC using test patterns as stimuli.*
   Many defects induce the same defective electrical behavior in the device; this complicates the process of diagnosis. In order to differentiate between defects, one has to come up with as many as possible fault equivalence classes and the corresponding measurement techniques for detecting these fault classes, which requires the full range of measurement techniques from voltage measurement to current measurement [Maly89, Soden89, Rodriguez91]. Using these means, one can deduce the defect characteristics, called *signature*, from the test data. Of course, a critical issue is the availability of meaningful test data.

5. *Interpretation of the results of testing by*
   a. *relating the results of testing to a list of faults,*
   b. *relating the list of faults to a list of defect candidates.*
   By comparing the actual tester data with the simulated test data it is possible to identify a list of defect candidates. The quality of the matching between test results and defect candidates will depend on the set of the simulated signatures and the matching between simulated signatures and actual signatures.

The beauty of this approach is that the product itself is now used as a defect monitor, and, as such, the methodology is really a process monitor testing method. Although the reasoning is simple, it will be clear that in practice it is difficult to implement this strategy. Computation of the signatures is a complicated process and it is the most important factor in achieving a diagnostic resolution. Further, activities such as test pattern generation for the various fault classes, measurement techniques such as I_{DDQ} monitoring, and statistical data analysis, are required to achieve the stated objectives. Process Monitoring Testing is currently not possible for large designs. It can, however, be applied to small design entities and to repetitive designs such as memories. The method has been successfully applied to large Static Random Access Memories. The results of this project made it possible to locate and predict a large number of defects on the 8k8 SRAM, while using the test results as the only source of information [Naik93].

It is shown in Chapter 3 that the possibility to fine-tune a test for a target fault model, with the capability of diagnosing the cause of the defect, is one of the backbones of the Macro Test concept. It is also shown that the limitation of the defect-oriented approaches with respect to circuit size is non-existent for Macro Test. An example of the use of the Inductive Fault Analysis technique, applied to a Static Random Access Memory, is given in Chapter 4.

3 Macro Test: A Framework for Testable IC Design

This chapter provides a descriptive introduction to the philosophy of *Macro Test* which is the concept at the very heart of this book. A clear understanding of what Macro Test is and the problems that it can solve allows the benefits of using this methodology to be fully appreciated. It is shown that Macro Test concepts need to be integrated into a design system. Some important integration issues, such as clearly described responsibilities, activities, and interfaces are discussed.

3.1 Introduction to the Macro Test Philosophy

There are various ways of reasoning towards a Macro Test approach. The first route is via the arguments of pertaining to test quality. The second route is via the arguments pertaining to modern IC design styles and the integration of a test methodology into such a design cycle. Both routes are described.

3.1.1 Macro Test driven by Quality Requirements

The twin objectives of zero defects and zero escapes, as described in Chapter 1, have serious consequences for test program generation. Both the customer and the supplier of a device are interested in the quality of the device. The zero escape driver is primarily meant for the customer. The driver is that no bad device passes a supplier's test and is shipped to a customer. As such, the tester acts as a filter; see Figure 3.1. An escape is a device which passes the test and is found by the customer not to conform to specifications.

A supplier is interested in the reasons for the failure of a device which does not pass the test. This knowledge is used to repair the device, to adapt the design, to improve the production process, or to improve the test program; see Figure 3.1. A zero defects approach implies that a supplier is trying to maximize on yield in order to maximize his profit and minimize the cost of producing the devices.

The notion of zero escapes is directly related to the quality of a test program as measured by its *coverage of defects*. The higher the quality, the lower the number of possible escapes, i.e. bad devices that pass a test. A way to indicate the quality of a test program is by stating the *fault coverage* of a test program with respect to a certain fault model.

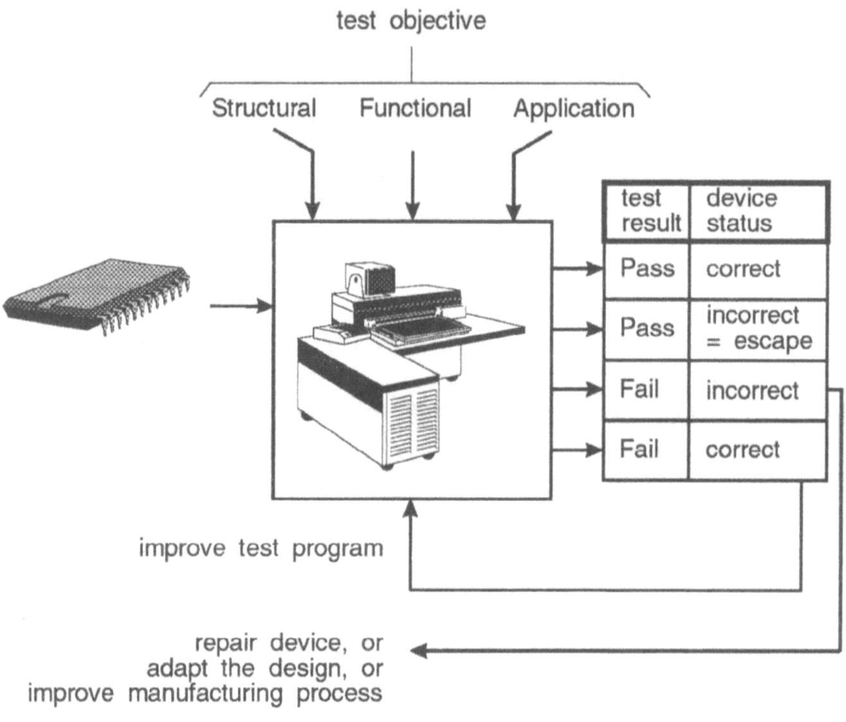

Figure 3.1. The tester acting as a filter.

As explained in Chapter 2, detailed studies are required to enable improvements to be made in the fault modeling process. Such studies enable realistic defect modeling and defect-oriented testing techniques. However, making use of defect-oriented techniques creates a layout dependency and technology dependency in the test generation process. Consequently, if different layout techniques are used on a single device in one single technology, then these parts should be tested differently. This requirement has resulted in the development of specific test techniques for SRAMs, DRAMs, ROMs, PLAs, array multipliers, etc. [Ligthart86, Maly86, Dekker88a, Stans89, Goor93, Sachdev93, Sas93].

However, the currently available defect-oriented test techniques are limited to small design entities; see Chapter 2. It is currently impossible by using these techniques to treat a large device which consists of millions of transistors as one single entity.

As a consequence, there is a need for a test methodology which is capable of creating a test program with excellent defect detection capabilities and defect analysis possibilities. Provided that the defect-oriented test techniques give the best results but are limited to small design entities, we can derive the requirements for this test methodology from a quality point of view as follows.

Quality 1. An ability to incorporate specific fault models / defect models and test techniques in different design entities.

Quality 2. An ability to generate test data based on these fault models and to execute the required test on these design entities.

3.1.2 Macro Test driven by IC Design Styles

A few observations have to be made with respect to IC design styles and methods and their respective consequences for a Test Methodology.

The first observation is that *embedded memories* and *module compilers* are being used in VLSI designs [Woudsma90]. This gives rise to problems associated with the mix of standard cell logic and non-standard logic in one single design. Most of the available *test pattern generation* (TPG) and *fault simulation* (FS) tools are based on a gate-level netlist representation of the design, and the tools cannot cope with embedded logic which cannot be described at a Boolean standard cell level. The problem becomes even worse in the case of module compilers and *silicon compilers* where a direct mapping from a high-level specification onto a layout is provided. A gate-level netlist may not be available for these cases which means that most of the generic TPG/FS tools may not apply to these kinds of designs. The use of complex libraries and module compilers forces a separation of tasks between 'library or module compiler' test development and test development for the total design compiled from these library entities.

The second observation is that in the focus on the design style the attention is paid to *architectural synthesis*. The trajectory from standard cell to layout is mostly automated. As soon as the design has been determined at the gate level, it is only a matter of weeks before the design is completed and can be submitted for manufacturing [Guterl84]. If the test tools can only be used when the gate-level netlist is completed then they are being applied at the bottleneck phase of product finishing. This is illustrated in Figure 3.2. To circumvent this situation, the testability aspects have to be taken into account early in the design phase, such that when the design phase has been finished, the test program has also been finished and verified. An additional benefit of a verified test program is that a verification, carried out by means of switch-level simulation, on a netlist which is extracted from the layout provides a thorough design verification.

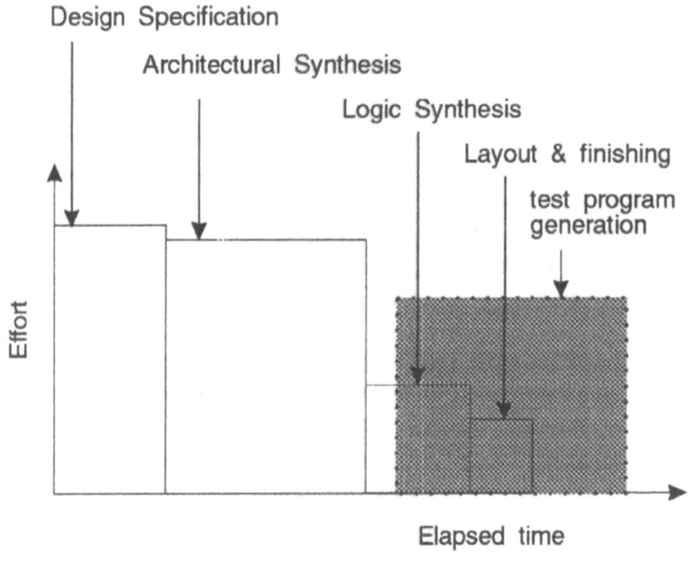

Figure 3.2. Test in the bottleneck of product finishing.

However, being early in the design cycle with test development means also that we do not deal with one single designer but with a *complete design team*. The design of complex VLSI devices is a *team effort* nowadays. Typically, the device is partitioned into *hierarchical sub-circuits* and the design of each sub-circuit becomes the responsibility of one or more members of the design team. The design style of each sub-circuit might be different (logic synthesis onto standard cells, sea-of-gates, full-custom, etc.) and, therefore, gradually, the total design has to be built up from the individually designed sub-circuits to form the complete device.

The *partitioning* of a device and the variety of design styles not only raise some interesting test challenges, but also supply the key to the provision of a good test solution [DasGupta84, Abadir85, Somenzi85, Beenker86, Murray88, Abadir89, Beenker89, Catthoor89, Leenstra90].

A third observation has to do with *time and money*. The time required to produce an IC is rapidly decreasing. As a consequence, the time needed for test program generation has to decrease as well, otherwise the whole design process would experience delay. Further, the price of a product is mainly determined by the market. Therefore, there is a limit to the amount of money which can be spent on a design and, consequently, on the total budget of expensive manhours. Today, the IC package is a dominant factor in determining the price of an IC. This is certainly the case for consumer-type of products. Every package has a maximum die size and a maximum number of pins. Consequently, the device has to fit into the prescribed die size, and considerable effort is spent to meet that target. This is one of the main reasons why most companies do not like to use silicon area or pins for testability provisions.

Another cost factor has to do with the use of automatic test equipment (ATE) for both prototype verification and production testing. Throughput requirements and memory impose limits on the available test time of a production tester. A reload of a tester memory is especially expensive.

With these observations as our basis, we can state the following requirements for a test methodology from a design point of view.

Design 1. Solve the testability issues early in a design cycle and not only at the gate level.

Design 2. Be flexible in the testability approach in order to be able to optimize all factors. However, the guarantee of a solution should always be available.

Design 3. Be able to work incrementally with different design teams and when the design is being assembled continue the activity based on the gathered information.

3.1.3 The Macro Test Concepts

Employing the set of test methodology requirements, we now derive a test methodology which meets these requirements. We have called this test methodology *Macro Test* [Beenker86, Beenker89, Beenker92].

The Macro Test methodology is based on *partitioning* a device into testable blocks, called *leaf-macros*, and then providing *test access* to every leaf-macro from the device pins; see Figure 3.3. This partitioning, which will normally map closely onto the logic design partitioning, allows the use of *macro-specific* test strategies and defect/fault models with their associated test pattern generation tools.

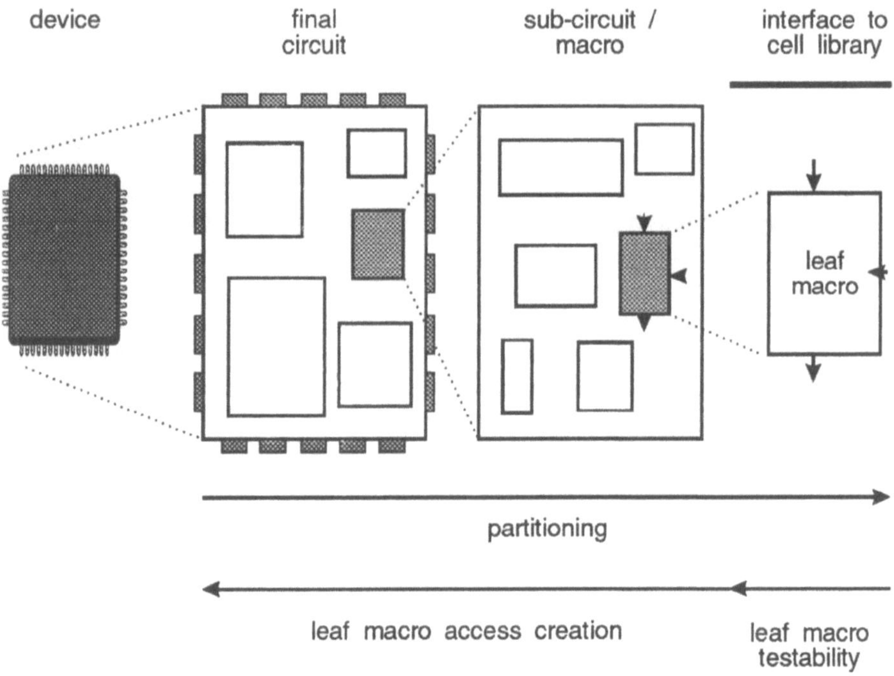

Figure 3.3. The hierarchical Macro Test concept.

Figure 3.3 shows both the design partitioning and the testing concepts. With regard to the partitioning, the entire device is divided into *macros*, which are then repeatedly divided into smaller and smaller macros. The lowest level macro is known as a leaf-macro and is a direct interface to a cell library or to a synthesis program.

With regard to the design-related requirements, by mapping the Macro Test partitioning onto the design partitioning, the incremental process of Macro Testing makes it possible to treat each design partition as a separate test task. This means that it is not necessary to wait until the design is assembled as a complete gate-level netlist, for example, before test related activities can commence. The test related work can start as soon as the design specification and architectural definition start. Consequently, the Macro Test technique allows an *early start* to be made on *testability as part of the design phase*. This leads to the removal of the usual bottleneck of test program generation at the end of the design stage.

The underlying principle of Macro Test is that each leaf-macro is designed to be testable. According to the definition of testability, given in Chapter 1, this implies that test patterns exist, or the means for creating acceptable test patterns exist, and the *test protocol*, which describes *how to test* each leaf-macro as a stand-alone entity, is known; see Figure 3.4.

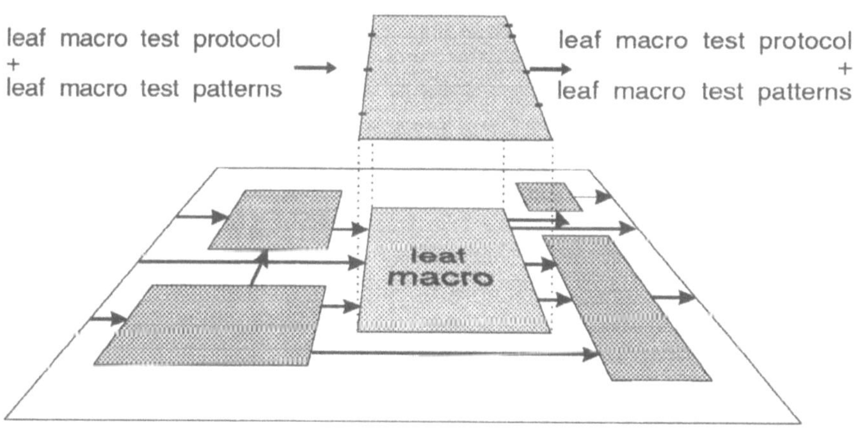

Figure 3.4. A leaf-macro as a stand-alone entity.

Then, by creating access to each leaf-macro through each level of the design hierarchy up to the top level, it is guaranteed that the whole design will be testable from the device pins. The beauty of this approach is that most of the test related work can be done in parallel with and even integrated with the logic design activities. This is because it is *test strategy* that is being dealt with and this is *independent of any actual test data.*

The requirement to provide access to every leaf-macro makes it mandatory to integrate testability into the design cycle because such access is a design issue [Abadir85, Abadir89, Bouwman92a]. How access is generated is not important as long as access is available in order to enable the required test procedure for every leaf-macro to be successfully executed, e.g., a RAM self-test or the use of scan for testing an ALU [Roth89]. The starting point for providing such access is a protocol which considers each leaf-macro as if it were a stand-alone unit. This leaf-macro specific protocol is called the ***initial test plan***. The initial test plan describes the *application protocol* of test pattern data to and from the macro described in terms of actions performed on leaf-macro ports. This initial test plan is applied for *each* test pattern in the leaf-macro test pattern set.

As an example of an initial test plan, consider Figure 3.5. A leaf-macro m1 is given. It has one scan chain which consists of five scan cells. The mode of operation of these scan cells is controlled by two test control signals, scan_ctr1 and scan_ctr2. In addition to the scan chain input port and scan chain output port, leaf-macro m1 has four input ports and one output port. The initial test plan for m1 is scan-based and is indicated in Figure 3.5. For every test pattern of m1, the input data for the scan cells is shifted in via the scan chain input port scan_in. The input data for the other input ports is applied at the moment the shifting process is finished. The response is observed at output port out1 and via a shifting process at the scan chain output port scan_out. The test protocol is controlled by a value sequence applied to the two test control signals. The example uses a language which is part of the Macro Test supporting Panther Expert 1.0 software [Beenker94]. We assume that with the added semantics, the reader will easily understand the syntax of this language. Note that the initial test plan does not contain any test pattern value information. It does contain a reference to test pattern data via statements such as 'in1<5> = in1_pat'. The statement refers test pattern data for port in1 at clock cycle 5 to an identifier in1_pat. The identifier also has to exist in the actual test pattern set of m1. In this set the identifier will refer to the actual data.

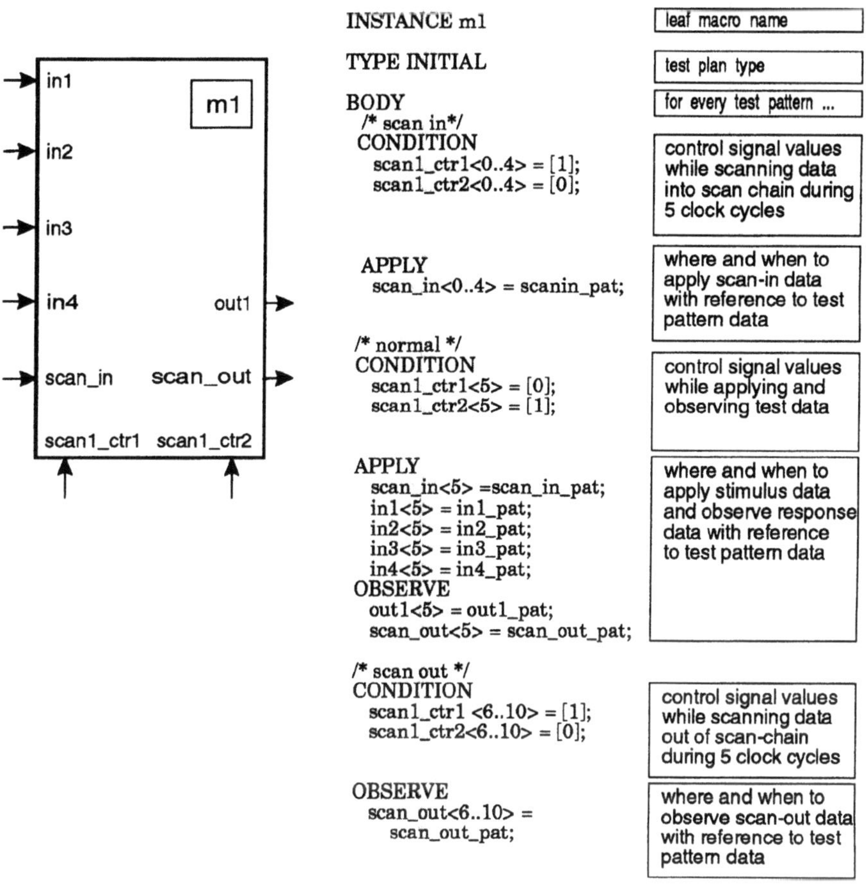

Figure 3.5. Example of a leaf-macro initial test plan.

An initial test plan is transformed gradually into what is known as a *generated test plan.* A generated test plan is the access protocol which specifies *how* a leaf-macro is to be tested *from the pins of the device*; see Figure 3.6. The transformation process, called *test plan generation*, is a design process whereby *access is created* from the device pins to the leaf-macro ports such that the initial test plan can be executed from the device pins. The example given in Figure 3.6 is a continuation of the example in Figure 3.5.

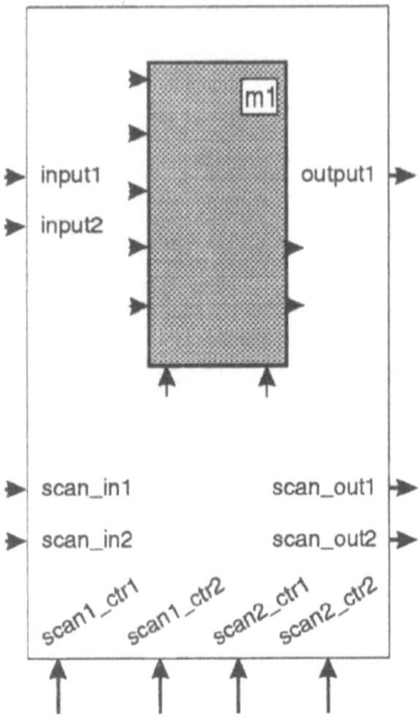


```
INSTANCE m1

TYPE GENERATED

BODY
 /* scan in*/
 CONDITION
   scan1_ctr1<0..4> = [1];
   scan1_ctr2<0..4> = [0];
   scan2_ctr1<0..4> = [1];
   scan2_ctr2<0..4> = [1];
  APPLY
   scan_in1<0..4> = scanin_pat;
   scan_in2<0..1> = [XX];
   scan_in2<2> = input2_pat;
   scan_in2<3..4> = [X];

 /* hold */
  CONDITION
   scan1_ctr1<5> = [0];
   scan1_ctr2<5> = [0];
   scan2_ctr1<5> = [0];
   scan2_ctr2<5> = [1];

 /* normal */
 CONDITION
   scan1_ctr1<6> = [0];
   scan1_ctr2<6> = [1];
   scan2_ctr1<6> = [0];
   scan2_ctr2<6> = [0];
 APPLY
   scan_in1<6> =scan_in_pat;
   input1<6> = m1_in1_pat;
   input2<6> = [0];
 OBSERVE
   output1<6> = out1_pat;
   scan_out1<6> = scan_out_pat;

 /* scan out */
 CONDITION
   scan1_ctr1 <7..11> = [1];
   scan1_ctr2<7..11> = [0];
   scan2_ctr1<7..11> = [1];
   scan2_ctr2<7..11> = [1];
 OBSERVE
   scan_out1<7..11> =
      scan_out_pat;
   scan_out2<7..11> =
      [UUUUU];
```

leaf macro name

test plan type

for every test pattern ...

control signal values while scanning data into scan chains during 5 clock cycles

where and when to apply scan-in data with reference to test pattern data

control signal values while holding values in scan cells fixed for 1 clock cycle

control signal values while applying and observing test data

where and when to apply and observe data with reference to test pattern data

control signal values while scanning data out of scan-chains during 5 clock cycles

where and when to observe scan-out data with reference to test pattern data

Figure 3.6. Example of a leaf-macro generated test plan.

Access to the ports of leaf-macro m1 is obtained by making use of two scan chains and by making use of existing functional data paths. All actions stated in the generated test plan are defined on the device ports. In the example, actions can be recognized such as applying and observing of data, and applying a value to a conditional signal to enable the test data transport. The reference to the test pattern data is identical to the previous example. The conditional statements of a generated test plan separately constitutes the *conditional test plan.*

The designer can make trade-offs between the possible ways of providing access to each leaf-macro - for example, by making use of existing functional data paths or by creating additional data paths and control signals specifically for test data transport. It is this flexibility that can be used to minimize any additional silicon area that may be needed to incorporate circuitry for testability purposes. Using the Macro Test approach typically limits the silicon 'overhead' required for testability purposes to less than 5% of the total chip area [Kuban84, Gelsinger87, Mehtani90, Bouwman92a].

In fact, the misconception that testability is an overhead should really be eliminated when testability is integrated in the design cycle. At this point testability becomes simply yet another design constraint.

Having created a generated test plan for each leaf-macro, the next step is to produce a *leaf-macro test specification*. This is done by merging the leaf-macro test patterns with their corresponding generated test plan; see Figure 3.7. This step is called *test pattern insertion*. The leaf-macro test patterns have been generated by tools, such as test pattern generators and fault simulators, or manually.

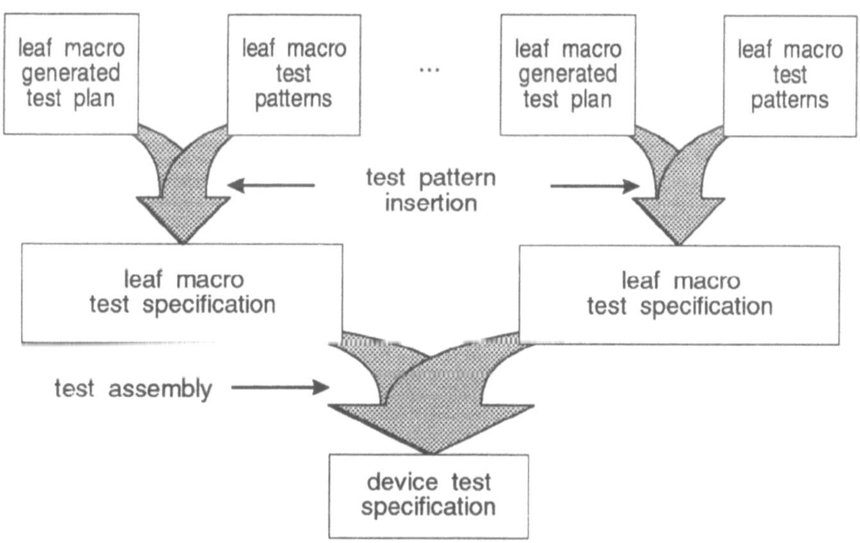

Figure 3.7. Test pattern insertion and test assembly.

A complete *device test specification* can be created by simply concatenating the leaf-macro test specifications or by merging them in order to reduce test time; see Chapter 7. The merging process of leaf-macro test specifications into a device test specification is called *test assembly*; see Figure 3.7.

By bringing testability issues into the design specification phase, a division is made between the design actions performed to produce a testable design and, given a testable design, the data manipulation actions necessary to produce sets of test patterns. It is common practice to gather these two phases together using the term *test synthesis*. However, to emphasize this *separation of concerns*, we prefer to use a slightly different terminology in this book. The first set of actions is referred to as *testability synthesis*. The second set of actions is known as *test synthesis*. This separation of concerns in design-related activities and data-related activities is one of the most essential items of Macro Test.

3.1.4 Macro Definition

An essential issue in Macro Test is the definition of a macro in the design hierarchy. A generic definition of a macro is difficult to provide. One could compare the complexity of a macro definition with that of an architectural specification of a device at the moment when a high-level functional specification of the design is available. However, such a definition cannot be immediately obtained. Design strategies iterate on the architectural level and high-level synthesis techniques and architectural-driven silicon compilers are used for this purpose. It is during the architectural specification discussion in the design phase that all macros are defined. The following general statements can be made.

1. To identify macros, we follow the design hierarchy closely. This procedure eases the integration of the testability synthesis activities in design. It eases the communication between the various design teams and the person responsible for the testability synthesis activities. Consequently, a macro is normally a sub-circuit of the design; see Figure 3.3. The design is considered as an hierarchical structure of macros. The lowest level macro is a leaf-macro.
2. For every leaf-macro there has to be a suitable defect/fault model with the associated test pattern generator/fault simulator.

Mapping an hierarchical structure of design primitives onto a layout does not necessarily require that the layout reflects the design hierarchy. During layout optimization the hierarchy might get lost. The same holds for a macro definition. The final device test specification is an assembly of the set of leaf-macro test specifications. The assembly can be done by simply concatenating the various leaf-macro test specifications and testing leaf-macro per leaf-macro. Another possibility is to optimize with respect to test time and to merge the leaf-macro specifications, thus obtaining a device test specification with minimal test time application [Bouwmeestr93]. As shown in Chapter 7, this method yields a device test specification where there is no clear mapping between a part of a device test specification and a single leaf-macro. Each part of this assembled device test specification will test a section of the hardware in one or more leaf-macros. We, therefore, can extend the concept of a macro to a *virtual macro*. The set of virtual leaf-macros covers the complete design.

Henceforth, the term leaf-macro is used to identify a design entity which has its own specified defect/fault model, test pattern generation/fault simulation method, and initial test plan.

3.2 Testability Synthesis within the Macro Test Concept

When using the Macro Test concept, a designer has a framework that guides him through the process of dealing with the testability issues associated with complex devices. The process of testability synthesis within a Macro Test concept is as follows.

For each leaf-macro, the designer specifies an initial test plan. The initial test plan describes the application protocol of test pattern data to and from the macro, assuming it is a stand-alone unit. An example of an initial test plan is given in Figure 3.5. The initial test plan does not depend on the test data. It contains *references* to test data and a specification of where to apply and observe test data. An initial test plan guarantees the same protocol is used when each test pattern is applied. A leaf-macro may have multiple initial test plans. For example, an embedded memory might be tested with Built-In Self-Test or with an externally provided test algorithm; see Chapter 4. Providing both possibilities improves the diagnostic capabilities of the test program.

With the initial test plan as a basis, the next step is to analyze the accessibility of the macro from the terminals of the sub-circuit in which this macro is embedded; see Figure 3.3. This process continues until the device pins have been reached or it is determined that no access path exists. This accessibility analysis requires data on global control signals, such as global reset values, and test control signals which enable the flow of test data. This data information is called a *conditional test plan*. An example of a conditional test plan is given in Figure 3.8.

The information in conditional test plans is derived from an analysis of the device architecture, the clocking schemes, the possibilities to initialize some or all modules, and any other factors which may influence data flow and communication protocols within the design. As with initial test plans, conditional test plans relate to leaf-macros and are independent of test data. Of course, a generated test plan for leaf-macro, m1 say, contains the conditional test plan of leaf-macro m1. This can be seen from the examples given in Figure 3.6 and Figure 3.8. Note from Figure 3.6 that the actions defined in a conditional test plan are defined on the device ports. It will depend on the actual tools supporting the Macro Test concepts how the conditional test plan information is gathered. The information can either be supplied manually or it can be extracted automatically from a design library.

The existing functionality of other macros in the device is used to obtain access to and from each target macro. If access is not possible then it must be provided by the inclusion of scan elements or by other design modifications. It should be noted that there is also a second reason to go for a design modification for improving access. The analysis of resources used for transporting test data may lead to the conclusion that design modifications can further reduce test time. This is further detailed in Chapter 7. The analysis of resources results in an optimal implementation of testability: optimal with respect to the silicon area used for testability, test program run time, pinning used for testability, influence on overall device performance, etc. The data representing the conditional settings for a data path through macros is known as *transfer information.*

An example of transfer information is given in Figure 3.9. In this example, transfer information is specified for macro m1. Given a condition value setting for the ports scan_ctrl and scan_ctr2, data will be 'transferred' from the scan_in port to the scan_out port in 5 clock cycles. This transfer information is referred to as 'inside_scan_info'. An extensive usage of transfer information is described in [Murray88, Roth89, Marinissen91].

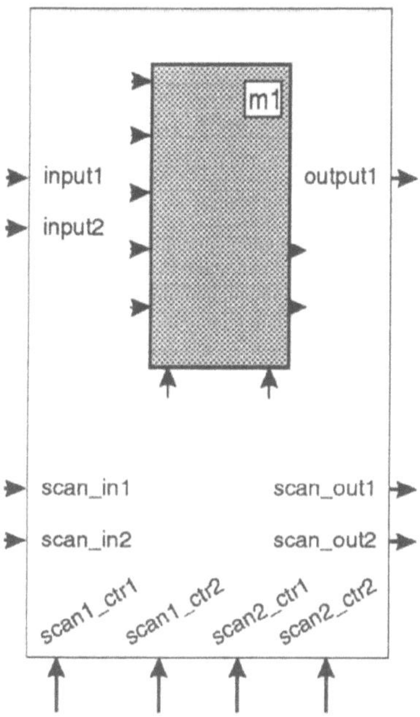


```
INSTANCE m1

TYPE CONDITIONAL

BODY
 /* scan in*/
 CONDITION
   scan1_ctr1<0..4> = [1];
   scan1_ctr2<0..4> = [0];
   scan2_ctr1<0..4> = [1];
   scan2_ctr2<0..4> = [1];

 /* hold */
 CONDITION
   scan1_ctr1<5> = [0];
   scan1_ctr2<5> = [0];
   scan2_ctr1<5> = [0];
   scan2_ctr2<5> = [1];

 /* normal */
 CONDITION
   scan1_ctr1<6> = [0];
   scan1_ctr2<6> = [1];
   scan2_ctr1<6> = [0];
   scan2_ctr2<6> = [0];

 /* scan out */
 CONDITION
   scan1_ctrl <7..11> = [1];
   scan1_ctr2<7..11> = [0];
   scan2_ctr1<7..11> = [1];
   scan2_ctr2<7..11> = [1];
```

leaf macro name

test plan type

for every test pattern ...

control signal values while scanning data into scan chains during 5 clock cycles

control signal values while holding values in scan cells fixed for 1 clock cycle

control signal values while applying and observing test data

control signal values while scanning data out of scan-chains during 5 clock cycles

Figure 3.8. Example of a leaf-macro conditional test plan

```
INSTANCE m1

PROPERTY inside_scan_info
IF scan_ctr1<0..4> = [1];
   scan_ctr2<0..4> = [0];
THEN scan_in TRANSFERS scan_out;
```

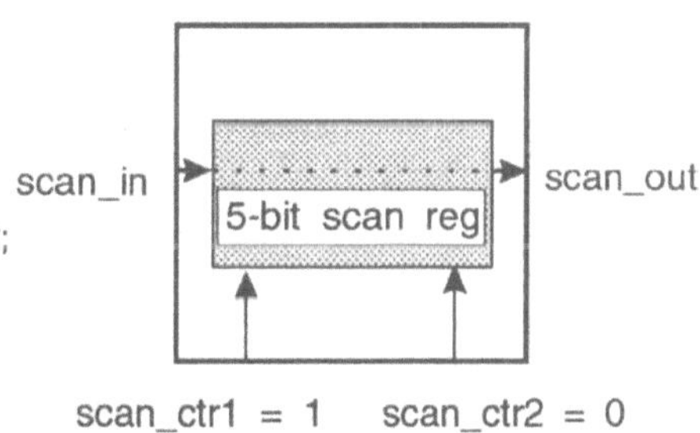

Figure 3.9. Example of transfer information.

The resulting access protocol to each Macro from and to the device pins is the generated test plan. In effect, the generated test plan is the initial test plan transported to the boundary of the device.

In cases where there is no access to a leaf-macro, the conclusion is that that part of the device is not testable. In this case, a designer has to modify his design such that it becomes testable. Whenever such testability hardware is added to improve testability, additional test control signals are introduced which control the flow of data during both the test mode and normal mode of operation through this additional testability hardware, e.g., scan control signals. Normally, the test control signals are provided via dedicated device pins. If the number of available pins is restricted, logic synthesis techniques can be used to synthesise a test control structure into the design to coordinate the execution of the generated test plans. With finite state machine specifications for this test control logic as a basis, it can then be synthesized by logic synthesis tools. The test control logic is called a *Test Control Block* (TCB) and it has one pin controlling the finite state machine; see Figure 3.10 and Chapter 6.

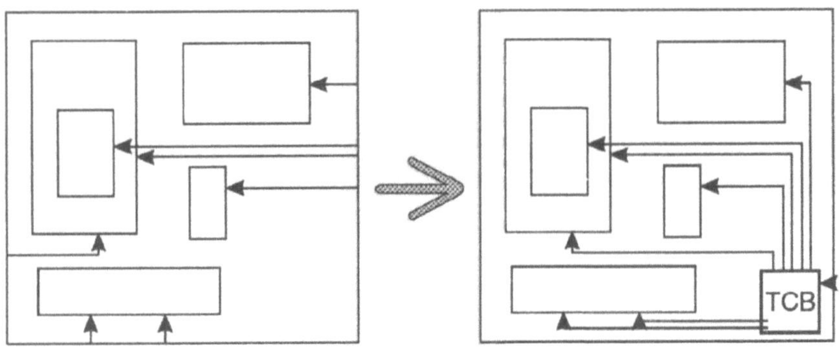

Figure 3.10. Function of a Test Control Block.

Testability Hardware which often resides in a netlist may consist of the following elements.

* Scan cells and scan chain routing inside macros and between macros; see Chapter 5.
* Test Control Block. The Test Control Block specification is given in the form of a finite state machine description. The finite state machine can be minimized with respect to the number of gates used in the next-state logic and globally with respect to the number of states; see Chapter 6.
* Boundary Scan logic conforming to the ANSI IEEE 1149.1 standard [IEEE90].

On completion of the testability synthesis, the device netlist is passed to tools such as test pattern generators and fault simulators for the generation of the test patterns, and to the design environment for design finishing activities, such as bond-pad placement and routing and final design verification.

3.3 Integration of Macro Test into a Design & Test flow

3.3.1 The Evaluation Plan

By clearly defining the testability synthesis phase to be part of an IC design process, we have integrated the testability issues into a design cycle. As stated, testability synthesis is bound to some requirements, limitations and boundary conditions, such as the maximum die size, required performance of the device, maximum number of pinning and required test quality. During the testability synthesis phase, a designer is guided in the task of producing a testable device, which means a device with a corresponding test program which meets all requirements and boundary conditions. To be able to cope with these difficulties, it is required that the responsibilities and targets are clear and that testability has become part of the initial design specification.

To make this clearly visible, one may introduce the notion of an *evaluation plan*, which summarizes all test-related matters of a design specification [Beenker94]. The purpose of the evaluation plan is for a target design to define the starting points, constraints and agreements on all test issues. The main discussion is to determine what will be tested and how it will be tested. It is not necessary to describe the tests in detail, only the information required to reach the aims is necessary.

The big advantage of the evaluation plan is that testability issues are *visible* to the project team and that *responsibilities* are assigned so that the discussion is open from the beginning and the knowledge of both test engineers and design engineers is combined. Thus, the implementation of the testability strategy becomes optimal for the target device and the boundary conditions can be met. Further, a discussion about 'overhead' is no longer relevant. Just as in normal design practices, the amount of silicon and number of pins used for the testability functionality is minimized.

The evaluation plan must be available from the start of a project and be signed by all responsible persons. During the project, the evaluation plan is detailed further.

The global contents of the evaluation plan are as follows.

The Device
This section of the evaluation plan provides a global description of the device in order to understand its basic functionality and that of the testability aspects of the device, and test- department related issues.

The testability aspects cover the following items. They specify which software is going to be used to support the testability synthesis and test synthesis activities, which design-for-testability methodology will be used and in which way the results will be guaranteed for correctness. They define the pins for testing, e.g., test control, scan-in, scan-out, and test-clock, and which parts of the device are to be covered by the test specification. Clocks are specified, the source of the clock signals, relationships between test and system clocks and between clocks when there is a multi-clock design, etc.

The test department specifies the type of tester required for the device test and the kind of handler required for the device or package. It further specifies the test language required for the tester and whether any problems are anticipated with the testing of the device.

The Tests
A specification of the test has to be provided for all parts of the design. The test pattern generation program to be used for pattern generation and the way in which the patterns are to be verified for correctness are specified. Further, an estimate of test data volume is given to get a first impression of target tester requirements on capacity and run time and to identify potential problems. The required fault coverage per macro with respect to the target fault model is specified, including the maximum test time for the total device.

The Hardware
With the evaluation plan sections on the device and the tests as its basis, the hardware evaluation plan specifies the tester to be used for prototype verification and the tester to be used for production testing. The load board is also specified and corresponding action is taken.

The Software

The difficulty in using software tools has often to do with the interfaces. During a design project, many different software tools are used, each with its own specific interface and input/output characteristics. To guarantee the availability of all input data and to perform some risk assessment, what software is to be used for which task should be clearly specified. The tasks to be considered are design of the device, testability synthesis, test synthesis, verification, and test program generation. All input and output data characteristics must be specified.

Responsibilities

In order to ensure that all activities are carried out and to clearly indicate the people to be contacted, all responsibilities must be assigned and the people involved must sign the responsibility allocation document. Responsibilities have to be assigned to various people for testability synthesis, test synthesis, verification of the test specification, software tools and interfaces, prototype verification testing and production testing.

3.3.2 Interfacing & Integrating

The evaluation plan defines the action required to produce a testable design. Performing the action involves a number of design and test activities. Design activities have to do with creating access to the leaf-macros: test activities have to do with creating the actual test program and performing a prototype verification test or production test. The integration into a design and test flow of the Macro Test methodology within the software tools requires clearly defined interfaces: interfaces to libraries and the design data bases for testability synthesis, and interfaces to tester languages and simulator languages for test synthesis.

Testability Synthesis Interfaces

The activities of a design team responsible for the testability synthesis activities are based on two information packages.

1. Information about the test procedure of a leaf-macro.
2. Information about the accessibility of leaf-macros.

The first information package specifies how a leaf-macro is tested as if it were a stand-alone unit. For example, it may specify a scan test of a leaf-macro, a Built-In Self-Test procedure, or a partial scan test. Normally, this information has to come from a library team or from the responsible test people. The information depends heavily on what the actual design system is able to support. For example, are specific module compilers available for producing a Built-In Self-test version of the target leaf-macro? Are scan cells available in the library? Is there a sequential test pattern generator supporting a partial scan version of the leaf-macro available?

Access to a leaf-macro is specified in three different categories: access for the test data, access for the test control signals and global control signals controlling the test data flow, and access for the clocks signals clocking the sequential elements of the device which influence the test data flow for a leaf-macro.

Access involves information about transfer information of design entities, for example transfer information of scan cells. Access for the test control signals and global control signals specifies the mode of operation of these signals and the test control sequences required to activate the test data flow. Access for the clock signals defines the operation of the clocks and the way they are activated. A lot of this information can be extracted automatically from the design library.

When access needs to be created to a leaf-macro, specific cells are required in a design library. Examples of such cells are scan cells, boundary scan cells, multiplexers, and bus-drivers and sensors. As a consequence, a generic coupling to a design library is required.

Finally, a netlist of the design is required in order to perform an access analysis for every leaf-macro.

Test Synthesis Interfaces

The output of the test synthesis phase is a test specification for each leaf-macro or for a set of leaf-macros. The test specification serves two needs. It is sent to a software simulator for verification of the generated test specification and it is the basis for the device test program and is therefore transformed into a tester language. The question is which language should be used for the test specification.

Many proprietary languages have been proposed and developed that describe and represent detailed digital behavior and each has its limitations and advantages. The problem is that most of these languages, in addition to being non standard, are often closely linked with either the target tester architecture or a specific simulator's capabilities. In tester languages there is usually a direct correspondence between language constructs and tester architecture features. Simulator stimulus languages are usually closely related to the timing and data requirements of a given simulator.

What is needed is a representation capable of explicitly representing and manipulating timing, data and format independently in a tester-independent and simulator-independent fashion. Such a representation is easy to understand, simple to manipulate, extensible, and compact [DenBeste86, Lunde89, Verhelst89, Baker93].

3.4 Summary of Essential Macro Test Items

The Macro Test concepts were derived as an answer to the requirements of a test methodology. The driving force behind the work on Macro Test was the desire to ensure the quality of a product, which was converted into zero escape, zero defect objectives. We have identified the following essential items in the Macro Test concepts.

1. A test is broken down into several parts. These parts are: the access, i.e., the physical paths to the stimulus, response, and control ports of the leaf-macro; the protocol to apply each test; and, finally, the test patterns. This approach results in a separation of areas of interest. A designer can concentrate on creating access to and from leaf-macros (testability synthesis) and a test engineer can concentrate on producing test data for leaf-macros (test synthesis).
2. Lower-level testability is preserved at the higher level. A distinction is made between test development for library modules and test development for hierarchical designs.
3. The partitioning into leaf-macros enables a division into manageable entities for the application of defect- oriented test techniques. This provides the ability to make use of leaf-macro specific fault models and the corresponding test pattern generation and fault simulation tools.

4. Macro Test allows an early start of testability in a design cycle, avoiding the bottleneck of test program generation during the design finishing phase.
5. If a corrective design action is required for creating leaf-macro access, then the designer is in full control of the design change and can make use of global knowledge of the design.

In general, the conclusion is that Macro Test makes the testability problem visible and manageable. This of course has its price also; we focus the attention on local issues of leaf-macros. This has two consequences.

1. Global aspects may be ignored. For example, scan chain routing can be done in an optimal way when all design details are known. Via the Macro Test approach, the scan chain routing is done hierarchically, which might yield a local optimum.
2. Interconnect between leaf-macros may not be tested with respect to the correct defect model of shorts and opens in interconnect. It can easily be seen that the interconnect between leaf-macros is tested according to the fault models used for the leaf-macros themselves. Interconnect acts either as a leaf-macro input port or as a leaf-macro output port. The leaf-macro fault model might not include the required fault model for interconnect. In practical applications of Macro Test two alternative techniques have been applied: either specific test patterns to test the interconnect on shorts and opens have been added, or an I_{DDQ} test to test the interconnect structure has been applied.

The following chapters of this book discuss the various Macro Test aspects. Chapter 4 presents some leaf-macro studies on fault models and test methods. In Chapters 5, 6, and 7 we discuss various aspects of leaf-macro access. These aspects include scan chain routing, Test Control Block generation, and optimization of the device test specification with respect to minimal test time application by accessing leaf-macros in parallel.

4 Examples of Leaf-Macro Test Techniques

Macro Test relies on the availability of an access protocol for every leaf-macro from the device pinning such that the corresponding leaf-macro test can be executed from the pins of the device. Initially, information is required on how to test the leaf-macro as a stand-alone entity. This information is described in the leaf-macro initial test plans. As explained in Chapter 3, an initial test plan is independent of the actual leaf macro test pattern set and is needed for the testability synthesis of the design. Hence, before starting the testability synthesis phase a study is needed on how to test every leaf-macro as a stand-alone unit. This study is guided by the capabilities of the module compilers and libraries as residing in a design system or by the capabilities of a logic synthesis tool used. For example, the design system may have RAM, ROM, and PLA module compilers, or may have a logic synthesis tool which is capable of generating a fully scannable leaf-macro or a self-testable leaf-macro.

The study how to test a leaf-macro is guided by the device test program objectives. The objectives can be stated in terms such as fault coverage, defect coverage, minimal set of test patterns to be applied, minimal test pattern generation effort, and minimal influence on performance of the leaf-macro. The more details of a specific leaf-macro are taken into account, the better the objectives can be achieved. An example of a leaf-macro detail which can be utilized is a possible repetitive or regular structure of a leaf-macro design. For example, a memory cell array consists of a repetitive structure of memory cells. An array multiplier is often implemented as a repetitive structure of full adder cells. A Programmable Logic Array (PLA) has a regular structure of input, product and output lines.

In this chapter, two examples of leaf-macro studies are given. Both examples concern a Static Random Access Memory (SRAM) [Dekker87, Dekker88a]. The objective of the first study is to obtain a leaf-macro test with excellent defect coverage capabilities, and minimal test application time. For this reason, an Inductive Fault Analysis study is performed on an SRAM design. The result of this study is an SRAM defect model and a test algorithm to test this SRAM for this defect model. As a second step, this test algorithm has been implemented in a self-test machine for the SRAM to provide a Built-In Self-Test option with an SRAM module compiler [Dekker87, Dekker88b, Rijk91].

Numerous leaf-macro studies with various objectives have been performed and can be found in literature. They include detailed fault modeling and test method studies of PLAs [Ostapko79, Ligthart86, Maly86], tests for array multipliers which make use of the regular structure of the multiplier [Friedman73, Shen83, Matterne87, Stans89], self-test for scannable and combinational designs [Konemann79, Hollmann90, Agrawal93a, Agrawal93b], and a range of test algorithms and fault models for memories [Hayes75, Suk81, Marinescu82, Papachristou85, Goor91].

4.1 Defect Modeling and Test Algorithm Development for Static Random Access Memories (SRAMs)

4.1.1 Introduction

Random Access Memories require special treatment for structural testing. Where in the past memories were fabricated with specialized processes, they are now integrated on one single substrate with other functional blocks using a process developed for standard logic. Memories have, by far, the most dense layout. As a result, the probability of defects in memory modules is higher than that of the normal logic which is not so densely laid out. This has led to new challenges in the design and test of embedded memories.

Their special test requirements have been recognized and addressed by several researchers. A wide variety of SRAM tests based on different fault models has been developed. A memory test is a specific sequence of write and read operations applied to each cell of the memory cell array. For example, a simple test for single stuck-at 0 faults requires a sequence of write 1, read 1 operations for every cell. For this reason we use the term '*memory test algorithm*' rather than 'memory test'.

Historical Overview

The first set of test algorithms that were published were mainly ad-hoc algorithms based on design considerations. No guarantee can be given about the defect detection capability of these algorithms. In 1977, Knaizuk and Hartman presented the ATS algorithm [Knaizuk77], covering all single and multiple stuck-at faults in a RAM. The assumption of a non-creative decoder structure, that is, that a single fault within the decoder does not create a new memory address to be accessed without also accessing the programmed address, and the rather difficult addressing order were disadvantages of this algorithm. Nair tackled these difficulties with the introduction of the MATS algorithms [Nair79]. At the same time Thatte and Abraham introduced the two-coupling fault model [Thatte77], i.e., a state or transition in one cell changes the state of the other cell. Besides covering all stuck-ats, the main achievement concerned the translation of decoder and read/write logic failures into equivalent memory cell faults. Marinescu [Marinescu82] and Suk and Reddy [Suk81] were inspired by this approach and proposed more efficient test algorithms. Papachristou and Sahgal [Papachristou85] published an even more extensive two-coupling algorithm. Hayes came in 1975 with the pattern-sensitive approach [Hayes75]. He stated that the most probable RAM faults are caused by possible influences between adjacent located memory cells. The approach assumes increased parasitical effects owing to the dense memory array.

A good overview of these algorithms can be found in [Veenstra88, Goor91].

Motivation

The evolution of Random Access Memory test algorithms is closely related to RAM defect model and fault model development. Fault models, such as the stuck-at fault model, the coupling fault model, and the pattern sensitive fault model have been presented and respective test algorithms have been developed. Most of the proposed fault models and test algorithms were hypotheses and rules of thumb and were not based on the actual manufacturing defects. Therefore, there was the need for an efficient and accurate test algorithm targeted on actual defects. As explained in Chapter 2, the Inductive Fault Analysis technique is appropriate for test algorithm development [Maly85, Shen85, Ferguson88a].

Objectives and Basic Approach

The main objective of our SRAM study was to show the feasibility of a fault model and test algorithm development based on actual device defects via an Inductive Fault Analysis approach as described in Chapter 2. The defects are modeled as local disturbances in the layout of an SRAM memory cell array and successively translated onto their electrical equivalent in the corresponding transistor diagram. The electrical behaviour of each defect is analyzed and classified, resulting in a fault model at SRAM cell level. To keep this analysis manageable in both time and complexity we restricted ourselves to single defects per memory cell.

As stated in Chapter 2, the following two defect categories are distinguished at layout level.

* *Global defects*, like too thick gate oxide or too thin poly silicon caused by process etching errors or oven temperature variations.
* *Spot defects*, like dust particles on the chip or the masks, scratches and gate oxide pinholes.

The impact of global defects covers a wide area. Hence, they are detected before structural testing by using simple structure measurements. A vast majority of defects that have to be detected during structural testing are caused by the local spot defects. For this reason, we only considered spot defects for our fault modeling purposes. Furthermore, we only considered catastrophic spot defects which result in a complete break or shorts in the circuit. Non-catastrophic defects such as nearly breaks and nearly shorts generally cause parametric faults [Syrzycki87, Bruls93].

Another objective of our work was to develop a test algorithm for SRAMs suitable for self-test applications. Therefore, effort has been spent to keep the test algorithm resulting from the Inductive Fault Analysis regular, symmetric, of linear complexity, and with a simple address order. Minimization of the test length was considered a second priority. The resulting test length can be optimized while maintaining the same fault coverage [Meershoek90, Goor93]. Furthermore, the resulting test algorithm can be used for both bit oriented and word oriented SRAMs.

4.1.2 Development of an SRAM Fault Model

An SRAM design is partitioned into three modules, the memory cell array, the address decoder and the read/write logic. The address decoder and read/write logic modules are further partitioned into registers, decoders, a driver, and a sense amplifier; see Figure 4.1. We will argue that for our purposes of defect modeling and test algorithm development, a complete SRAM can be considered as a leaf-macro.

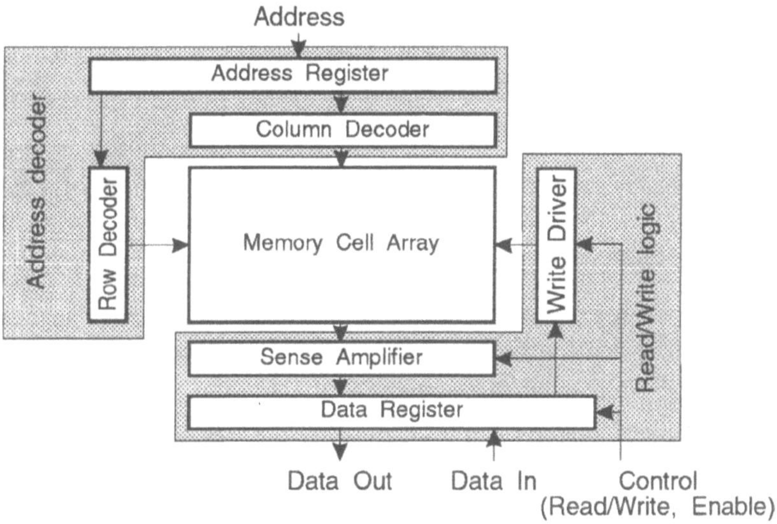

Figure 4.1. Functional SRAM model.

Initially, we considered each design module separately. The analysis of the memory cell array was done using the Inductive Fault Analysis technique. Due to the complex circuit structure of the address decoder and the read/write logic, and the limited capabilities of our IFA supporting software tools, we were not able to perform an identical analysis on the other two modules. For this reason, we assume a general functional fault model for both the address decoder and the read/write logic. It is shown that the considered faults can be viewed as memory cell array faults and that all faults in the memory cell array can be propagated through the read/write logic to the SRAM outputs. The conclusion is that the complete SRAM can be considered as a leaf-macro [Nair78, Muehldorf82, Goor91].

The following terminology is used to reason about *memory cell faults.*

Definition 4.1. A memory cell is said to be *stuck-at* if the logic value of the cell cannot be changed by any read or write operation on any cell in the memory array.

Definition 4.2. A memory cell is said to be *stuck-open* if it is not possible to read the contents of a cell nor to write data into the cell.

Definition 4.3. A memory cell with a *transition fault* fails to undergo at least one of the transitions $0 \rightarrow 1$ or $1 \rightarrow 0$. A transition $x \rightarrow -x$, $x \in \{0,1\}$, of a memory cell is a change of the cell contents from value x to value -x due to a write operation with value -x to the cell storing value x.

Definition 4.4. A memory cell i is said to be *state coupled* to another memory cell j if cell i is forced to a certain value x ($x \in \{0,1\}$) only if cell j has a given value y, $y \in \{0,1\}$. State coupling is an a-symmetric relation.

Definition 4.5. A memory cell i is said to have a *multiple access fault* to another memory cell j if a write operation with value x, $x \in \{0,1\}$, to cell i also forces a write operation to cell j with value x or -x. Multiple access is an a-symmetric relation.

Definition 4.6. There is a *data retention fault* in a cell if the cell fails to retain its logical value after some period of time.

It should be noted that these definitions have overlap. For example, a stuck-at fault is also a transition fault. For a further elaboration on the various faults in memory cells we refer to [Goor91, Goor93]

Inductive Fault Analysis of the Memory Cell Array

The layout of an 8k8 SRAM, designed and manufactured in 1986, was used as a vehicle to perform a fault model study. This memory is made in a double poly CMOS process with passive pull-up resistors. The pull-up resistors are 100 GΩ and made of high resistive poly silicon. The memory cell circuit structure is shown in Figure 4.2 and the memory cell layout is shown in Figure 4.3.

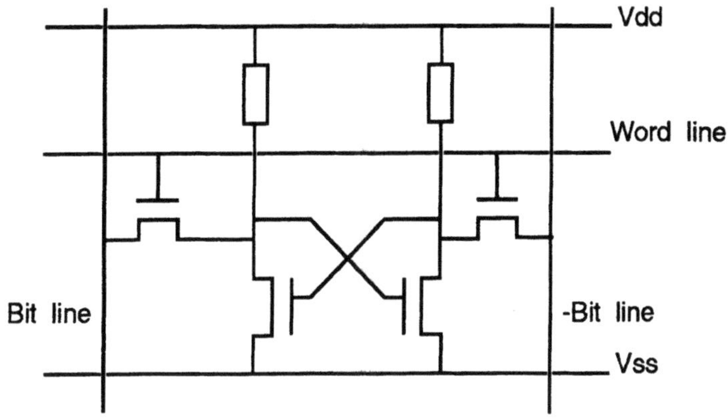

Figure 4.2. The 8k8 memory cell circuit.

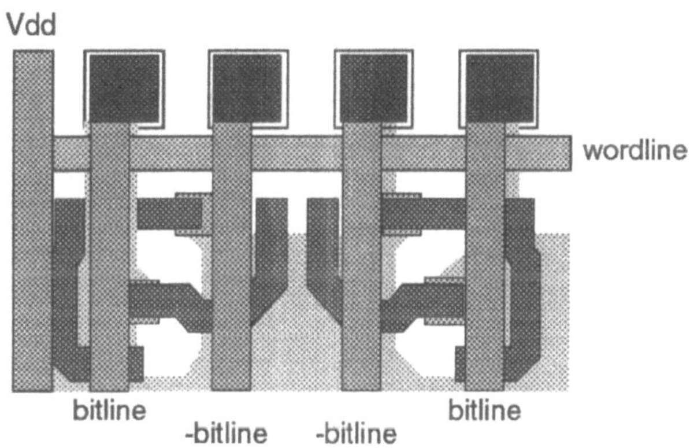

Figure 4.3. The 8k8 memory cell layout.

In order to model faults in the memory array, a physical defect model has been adapted. Catastrophic spot defects were modeled in the layout as extra or missing material. Notice that an SRAM cell array is a matrix of identical cells. The Inductive Fault Analysis was performed on a small repetitive section of the cell array. Defects were inserted per IC structure layer and each defect was analyzed separately on its influence. This resulted in five different layout-level spot defect types.

* Shorts between wires.
* Opens in wires.
* Newly created transistors.
* Missing contacts.
* Extra contacts.

The Inductive Fault Analysis was done in two steps. The first step was the translation of single spot defects in the layout to defects in the transistor circuit diagram. Many different layout defects result in the same electrical behaviour at the transistor diagram level. Those defects that were easily recognized as resulting in the same electrical behaviour were removed from the set of spot defects. A detailed analysis of the 8k8 SRAM resulted in a set of 60 spot defects in the layout that were analyzed in more detail. Figure 4.4 shows two examples of spot defects at the layout extracted to circuit level. Defect 1 is a missing piece of metal resulting in a broken bit line. Defect 2 is an additional piece of poly-1 which connects internal nodes and creates an extra pass transistor.

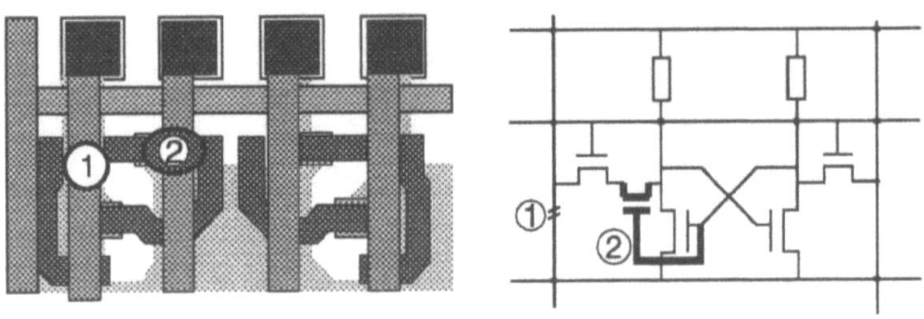

Figure 4.4. Examples of layout defects translated into transistor level circuit defects.
Defect 1: piece of missing metal.
Defect 2: piece of additional poly-1.

The second step was to classify defects at transistor level based on equivalent faulty memory cell behaviour. The result is a fault model at SRAM cell level, where the total set of defects could be divided over the following six fault classes.

1. A memory cell is *stuck-at 0* or *stuck-at 1*.
2. A memory cell is *stuck-open*.
3. A memory cell suffers from a *transition fault*.

4. A memory cell is *state coupled* to another cell.
5. There is a *multiple access fault* from one memory cell to a memory cell at another address.
6. A memory cell suffers from a *data retention fault* in one of its states. The retain time depends on the leak current to the substrate and the capacitance of the floating node. The retain time for thc 8k8 design can be up 100 msec.

The complete extraction of the fault model from the circuit defects is given in [Dekker87]. A few examples of the various fault classes are given in Figure 4.5.

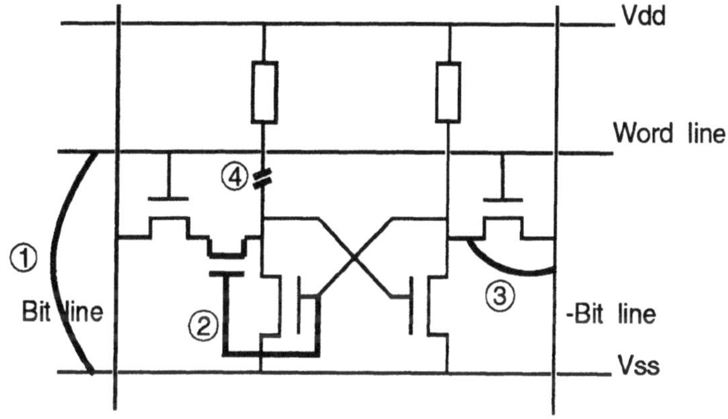

Figure 4.5. Examples of circuit defects for some fault classes.

Circuit defect 1 in Figure 4.5 connects a word line to V_{ss}. The cells in the word line are therefore not accessible. This defect causes all cells along the word line to be stuck-open. Circuit defect 2 is caused by a defect in the polysilicon layer covering a diffusion region. This defect may result in the creation of an extra pass transistor. This pass transistor is conducting if its gate is high, i.e., in one cell state. The cell is no longer accessible if the opposite information is written into the cell. This means that the cell suffers from a transition fault. Circuit defect 3 connects a cell node to a bit line. This defect introduces a state coupling fault. If the cell node is low (cell in state x), all other cells along the same bit line act as being stuck-at a certain value. If the cell node is high (cell in state -x) all other cells along the bit line function normally. Circuit defect 4 is a broken connection to a pull up resistor. This defect introduces a data retention fault. If the cell is not accessed, i.e. the pass transistors do not conduct, the cell node with the broken pull up resistor can be floating high or active low. If the node is floating high, the leakage current from the cell node to the substrate declines the voltage at the node. If the node voltage passes the threshold voltage V_{th} the data in the cell i is inverted. If the node is active low, the cell functions correctly. So the fault is only detectable at one state of the cell. For a detailed treatment of these defects we refer to [Dekker87].

The fault class importance is given by the probability of occurrence of a fault from this fault class. A simple metric has been used for determining this probability. The probability of a fault caused by a spot defect depends on the *critical area*. This is the chip area where a spot defect can damage the active part of the layout. The critical area depends on the dimensions of the spot defect and the topology of the layout. A detailed theory on critical areas can be found in [Pineda89, Pineda91].

For example, if two parallel metal wires are separated by 4μ, then a catastrophic short between these two can only be caused by a spot defect with a diameter of more than 4μ. The probability of occurrence of such a short is proportional to the length of the two wires. In general, the probability of occurrence of a defect of size δ is determined by the size of the critical area corresponding to δ [Pineda91]. We have taken a simplification of the critical area size by computing the length of wires in a memory cell which can either be shorted or broken by the defect. This length, which is dependent of the size of the defect, is called the *critical path length.*

The accumulated critical path lengths for each fault class and a range of spot defect dimensions have been calculated for the memory cell layout; see Figure 4.6. This leads to the following interesting conclusion about the importance of the single stuck-at fault model. Stuck-at faults indeed have the largest probability of occurrence, but the other fault classes are not negligible. It can be concluded from Figure 4.6 that roughly 50% of the occurring defects belong to the stuck-at fault class. This is a clear indication that the stuck-at fault model is insufficient for SRAMs. The critical path lengths, as presented in Figure 4.6, were based on the layout of the 8k8 memory cells. Different layouts will have different critical path lengths. The conclusions obtained from an Inductive Fault Analysis study are layout and process technology dependent. This means, that one cannot take over results without careful consideration. It means also that an Inductive Fault Analysis study has to be repeated after a change in technology or a change in the layout.

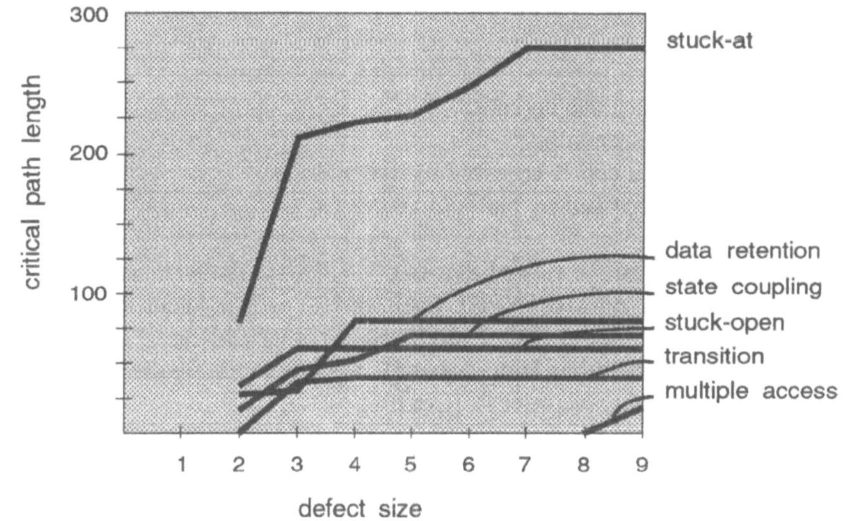

Figure 4.6. Critical path length per defect size accumulated for each fault class.

Analysis of the Address Decoder

The address decoder indicates the memory cell location in the memory cell array. The address is divided into row (or X) and column (or Y) addresses. The outputs of the X decoder drive the word lines in the memory cell array. The Y decoder selects one cell out of the columns connected to the word line.

The Inductive Fault Analysis technique complexity is mainly determined by the analysis of the electrical equivalent of the catastrophic defects. At the time of performing the study, the supporting IFA software tools were not capable of handling the complex circuit structures of both the address decoder and the read/write logic. Hence, an Inductive Fault Analysis, as performed on the memory cell array, was not feasible for these two circuit structures.

For this reason, a general functional fault model for the address decoder is assumed. It was presented by Nair, Thatte and Abraham [Nair78] and it assumes that the defects in the address decoder do not change the decoder into sequential logic.

The functional fault model for the address decoder consists of the following four fault classes.

1. An address accesses no cells.
2. A cell is not accessible.
3. More than one cell is accessed by one address.
4. One cell is accessed by multiple addresses.

Faults in the decoder can be viewed as memory cell array faults. Fault class 1 and 2 are equivalent to a stuck-open cell. Fault class 3 is equivalent to a multiple access fault from one cell to one or more cells at other addresses. Fault class 4 is a combination of the other fault classes. For a further elaboration on address decoder functional faults we refer to [Goor91].

Analysis of the Read/Write Logic

The read/write logic passes the data information from the I/O pins to the memory cell array and vice versa. During a read operation, the sense amplifier is connected to both bit lines. The sense amplifier determines the data value in the accessed cell from the voltage difference of the bit lines. During a write cycle, the new data is put on the bit lines by a write buffer.

We considered a word oriented SRAM with m bits per word. Similar to the address decoder only functional faults in the buses, sense amplifiers and write buffers were considered. This resulted in the following fault classes.

1. One or more of the m bits is stuck-at.
2. One or more of the m bits is stuck-open.
3. A pair of bits is state coupled.

All faults in the read/write logic can be regarded as faults in the memory cell array [Nair78]. Fault class 1 is equivalent to a set of stuck-at cells. Fault class 2 is equivalent to a number of stuck-open cells. Fault class 3 is equivalent to a state coupling fault between two cells at the same address.

4.1.3 Fault Propagation

Observation of the test data is done at the outputs of the read/write logic. Since all faults have been translated to faults in the memory cell array, fault detection is only assured when the read/write logic propagates all fault effects. When reading a value in a memory cell, the sense amplifier passes the proper defined logical value to the output pin. In case of a cell stuck-open fault the output value produced by the sense amplifier depends on its implementation.

If the sense amplifier is combinational, the sense amplifier passes a properly defined logical value to the output pin, i.e., always a 0 or always a 1. In this case, a cell stuck-open fault appears as a stuck-at fault and is detected. However, some designs of sense amplifiers include a data latch in the read path; see Figure 4.7. This data latch is used to broaden the read window of the RAM during normal operation.

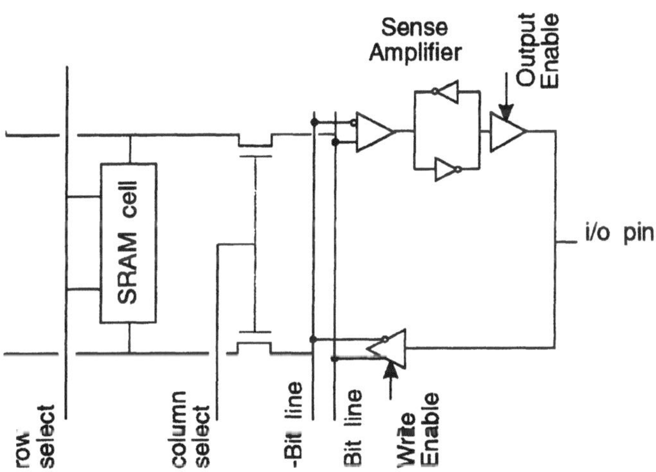

Figure 4.7. A latch in read/write logic.

The last read data of the memory is be stored in the data latch. In the case of a stuck-open cell, the latch might not be updated because the voltage difference between the bit lines is too small. In this case, the previous output value is passed to the output port. If this previous data is equal to the expected data then the stuck-open fault is not detected. This has to be taken into account in the SRAM test algorithm for applications where there is indeed a data latch in the read/write logic.

4.1.4 The SRAM Test Algorithm

Definition 4.7. A *march element* is a finite sequence of read and/or write operations applied consecutively to a cell in the memory array.

Definition 4.8. A *march test algorithm* consists of a sequence of march elements. After applying a march element to a given cell, the same march element is applied to the next cell until all cells have been treated. The address of the next cell is determined either in decreasing or in increasing addressing order. After applying a march element to each cell, the next march element of the march test algorithm is taken.

Definition 4.9. The *length* of a march test algorithm is defined as the number of march elements multiplied by the number of memory cells.

Below we present march test algorithms for both bit oriented and word oriented SRAMs with combinational or sequential read/write logic.

Bit Oriented SRAM with Combinational Read/Write Logic

A length 9N march test algorithm is presented, where N refers to the number of cells. In order to detect the data retention faults, a data retention test is added. The 9N algorithm is given in Figure 4.8. A Rd(0) instruction means read from the memory cell array and expect the logical 0 from the addressed cell. A Wr(0) instruction means write a logical 0 to the addressed cell. Similarly for Rd(1) and Wr(1). The address is indicated in the first column of the figure.

The proposed wait-time in the data retention test depends on the node capacitance and the leakage current in a memory cell. In the considered 8k8 a wait-time of 100 msec was estimated. Other cell designs or processes may result in other wait-times.

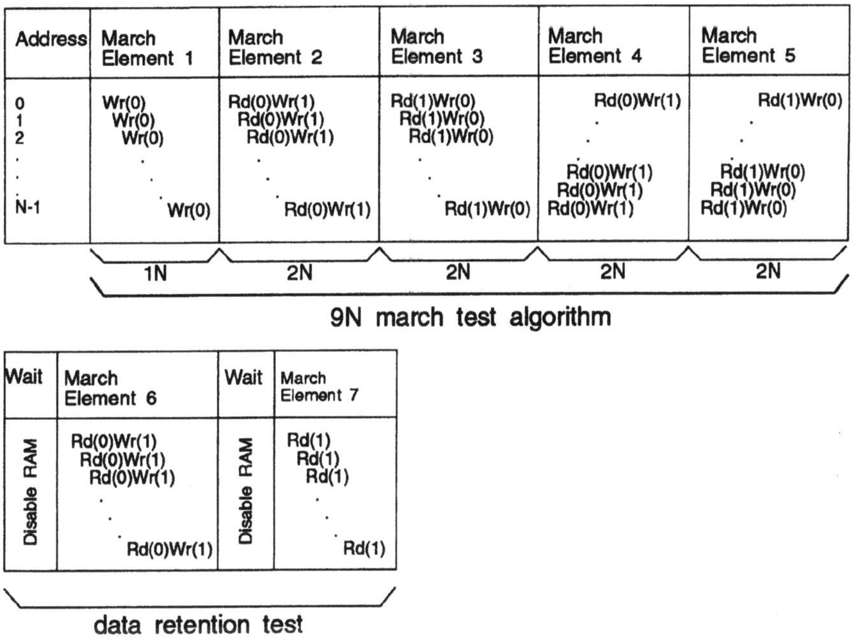

Figure 4.8. The 9N SRAM march test algorithm extended with a data retention test.

The following theorem states that the 9N test algorithm detects all faults of the fault model.

Theorem 4.1. The 9N march test algorithm detects all faults of the fault model for SRAMs with combinational read/write logic.

Proof

1. Each cell is checked for both states; Wr(1)Rd(1) in march element 2 and 3, and Wr(0)Rd(0) in march element 1 and 2. Each cell is checked for both transitions; Wr(1)Rd(1)Wr(0)Rd(0) in march elements 2, 3, and 4, and Wr(0)Rd(0)Wr(1)Rd(1) in march elements 1, 2, and 3. Hence, the test algorithm detects all stuck-at faults and all transition faults.
2. State coupling faults are covered if all four states of two arbitrary cells in the array are checked. The state diagram of Figure 4.9 shows that this is true with the 9N algorithm. Both cells are read for every state of the two cells. This is in particularly true for the two extreme cells at address 0 and address N-1.

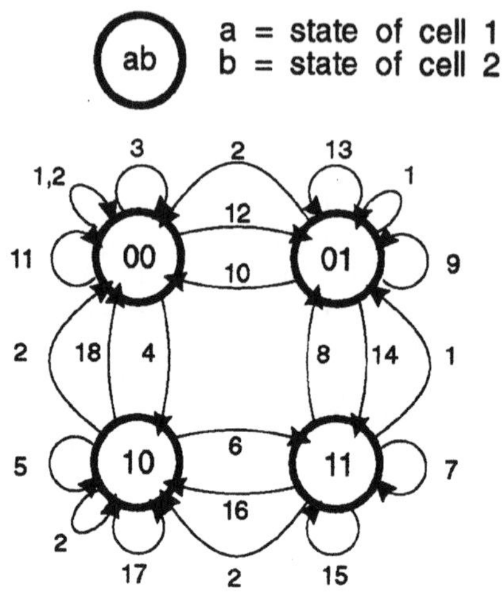

Instruction	State of cell 1 2
1 Wr(0) cell 1	
2 Wr(0) cell 2	0 0
3 Rd(0) cell 1	
4 Wr(1) cell 1	1 0
5 Rd(0) cell 2	
6 Wr(1) cell 2	1 1
7 Rd(1) cell 1	
8 Wr(0) cell 1	0 1
9 Rd(1) cell 2	
10 Wr(0) dell 2	0 0
11 Rd(0) cell 2	
12 Wr(1) cell 2	0 1
13 Rd(0) cell 1	
14 Wr(1) cell 1	1 1
15 Rd(1) cell 2	
16 Wr(0) cell 2	1 0
17 Rd(1) cell 1	
18 Wr(0) cell 1	0 0

Figure 4.9. Possible states of two arbitrary cells.

3. A multiple access fault from cell i to cell j occurs if cell i undergoes a transition to state x, $x \in \{0,1\}$, while cell j is in state x (-x). According to [Papachristou85], two march elements are sufficient to detect this fault for $N \geq j > i$: Rd(0)Wr(1)) and Rd(1)Wr(0) with an incrementing address order and initial memory state 0. For $0 \leq j < i$ the same march elements can be used with a decrementing address order. This is the case with the 9N march test algorithm.
4. Due to the combinational read/write logic, the stuck-open faults are equivalent to stuck-at faults and therefore detected. □

In a later stage it has been proved [Meershoek90, Goor93] that the 9N march test algorithm can be reduced to a 6N march test algorithm. This algorithm is shown in Figure 4.10.

Bit Oriented SRAM with Sequential Read/Write Logic

If the read/write logic includes a data latch, detection of stuck-open faults is not guaranteed by the 9N march test algorithm. This problem is solved by adding one extra read operation to each march element such that the expected read value is alternating high and low. With this addition it is ensured that the expected read value is always different from the latest read value. Thus each stuck-open fault is detected. The resulting 13N march test algorithm is given in Figure 4.11. The data retention test is added.

Word Oriented SRAM

A read or write operation for a word oriented SRAM involves reading or writing an entire word of data, called *data background.* The instructions in the march test algorithms must thus be redefined.

Wr(0) := Wr(<data background>),
Rd(0) := Rd(<data background>),
Wr(1) := Wr(<inverted data background>),
Rd(1) := Rd(<inverted data background>).

Word oriented SRAMs have the additional problem of state coupling faults between two bit positions in one memory cell, or between two bits in the read/write logic. To detect those faults all states of two arbitrary bit positions in one cell, i.e., 00, 01, 10, and 11, must be checked. This is only possible if several data backgrounds are used in the test algorithm. Detection of all coupling faults between two bit positions in one cell is not ensured by the 9N, 13N or any other test algorithm with only one data background. The proof of state coupling fault detection only holds for cells at different addresses. If m bits per word are used, a minimum of K data backgrounds is needed where,

$$K = \lceil \log_2(m) \rceil + 1,$$
$$\lceil x \rceil := \min \{ n \in \mathbf{N} \mid n \geq x \geq 0\},$$
$\mathbf{N}$ is the set of the natural numbers.

The proof of this statement can be found in [Dekker87]. As an example, we give the data backgrounds for m = 8, i.e., a byte oriented SRAM. According to the formula, K is 4.

Data background 1 : 00000000, Data background 2 : 01010101,
Data background 3 : 00110011, Data background 4 : 00001111.

The number of data backgrounds can be minimized if the topology of a particular design is known. The test algorithm is run with every data background separately. Finally, the data retention test is run only once. There is no need to run the data retention test for all data backgrounds.

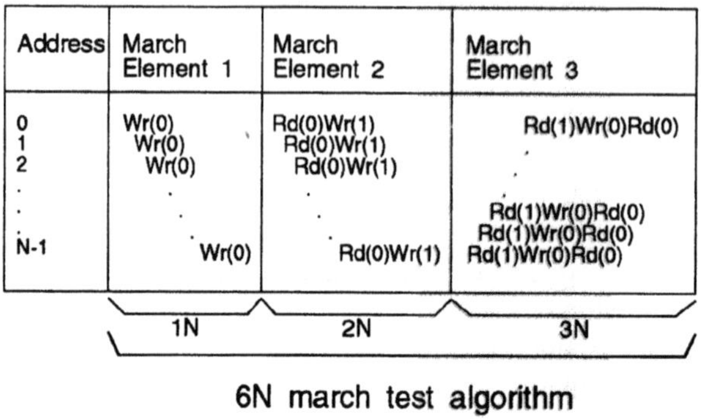

Figure 4.10. The 6N SRAM march test algorithm.

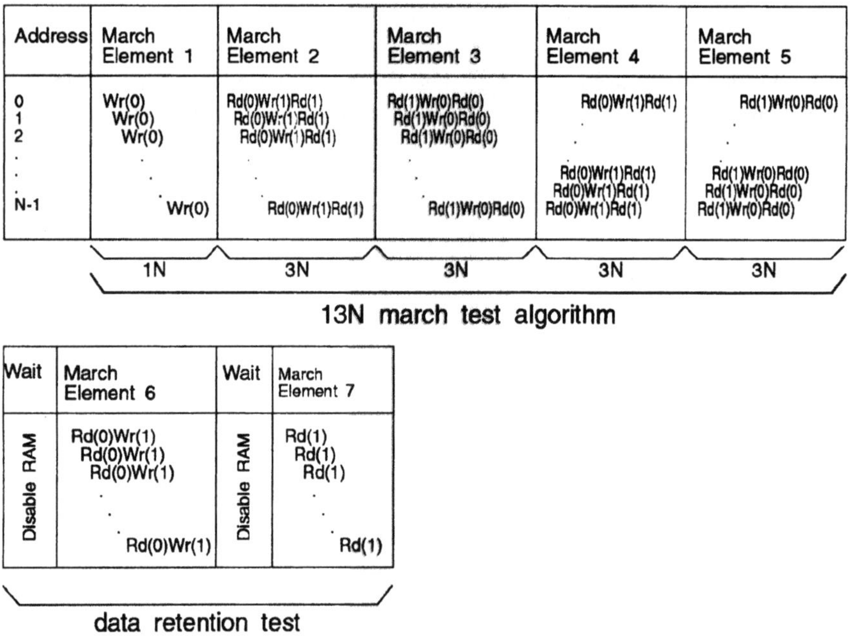

Figure 4.11. The 13N SRAM march test algorithm extended with a data retention test.

4.1.5 Practical Validation

The theoretical results obtained in the previous section have been validated by testing and analyzing a large number of the 8k8 SRAM devices. Defective memories have been evaluated and analyzed using microscopes and SEM techniques. In this way, the approach to derive a fault model, from an Inductive Fault Analysis layout study, was validated.

The Objectives

The validation objectives were twofold.

1. *Validation of the fault model.* Do the defects, as encountered in the SRAMs, indeed behave as described in the fault model ?
2. *Validation of the test algorithms.* How does our march test behave in comparison with test algorithms as proposed in literature ?

The Test Data Analysis

1192 8k8 SRAM devices were selected from 9 wafers from 3 different batches. These devices were tested with a set of 23 test algorithms on a Teradyne J389 memory tester. This set of algorithms consists of the 9N and 13N march test algorithms and a number of algorithms as proposed in literature. The choice of the latter algorithms was guided by a previously performed comparative study [Veenstra88]. Finally, various algorithm implementations, based on several addressing orders and data backgrounds, were used.

Executing a test algorithm on one device results in a number of detected faulty cells. This number is called the *bit error number*. It is dangerous however, to rely completely on the bit error number as a measure of test algorithm effectiveness. One defect in a decoder may result in a large bit error number. One defect in the memory array may result in a small bit error number. A second problem with the bit error number is the interpretation of the defects. The test algorithms have an overlap with respect to detection capabilities of the faults in our fault model. Correlations therefore exist between the different bit error numbers of the various test algorithms.

The bit error number can vary strongly from algorithm to algorithm and from device to device. We have translated defects in the decoders and the read/write logic into equivalent faults in the memory array. Hence, it is possible that, although for various devices the bit error number is completely different, the actual fault class is the same.

In order to find these correlations, the *behaviour* of the test algorithms over one device and the *behaviour* of the test set over the total number of devices has been computed. The behaviour of the test algorithms over one device is described by using the notion of a *signature*. The behaviour of the test sets over the total number of devices is described by the notion of a *cluster*. The signatures were also used to measure the effectiveness of the test algorithms.

For each device, each test algorithm is assigned a signature bit with values 0, 1 or 2. This bit is a function of the bit error number of the corresponding test algorithm. The value of the signature bit is determined according to the following step function SB.

Let BN be the set of bit error numbers for one device, $|BN| = 23$, and $b \in SB$. Let m be the mean value of the bit error number of BN and r the standard deviation of the bit error numbers of BN.

$$SB : BN \rightarrow \{0,1,2\},$$
$$SB(b) := \begin{cases} 0, & 0 \leq b < m - r/5; \\ 1, & m - r/5 < b \leq m + r/5; \\ 2, & b > m + r/5. \end{cases}$$

The decision factor of r/5 is chosen rather arbitrarily and resulted from a discussion with persons experienced in statistical test data analysis.

This process resulted in a 23 bit signature per device which may be considered as a fault-class representative.

Some of the most frequently occurring signatures differed at one or two signature bits. The differences in these signatures were caused by random defects and multiple faults on a device. To obtain the main fault classes the signatures were clustered using a statistical clustering algorithm, similar to the isodata algorithm developed by Tou and Conzalez [Tou74]. A cluster contains signatures that only show minor differences. The hypothesis was that each cluster could be characterized by a single fault class.

Validation of the Fault Model

Our analysis resulted into 15 clusters; see Table 4.1. Several devices from each cluster were analyzed in more detail using a light microscope and a SEM. Defects in devices from the same cluster indeed all suffered from the same fault class. The result of this analysis is shown in Table 4.1. Notice, that most of the theoretically derived fault classes are indeed important and were encountered in practice. Only transition faults were hardly ever found and multiple access faults were occurring regularly because of faults in the decoder. Examples of spot defects in devices from several clusters can be found in [Dekker87, Dekker88a].

cluster	#devices	observed fault class
0	712	cell stuck-at & total chip failure
1	169	cell stuck-open
2	18	multiple access
3	9	state coupling
4	8	?
5	5	?
6	26	cell data retention
7	3	?
...	...	...
13	3	?
14	2	?

Table 4.1. Resulting clusters with the observed fault classes.

About 10% of the analyzed defects showed unexplained behaviour. In a later stage of research activities it has been proven that the march test algorithms detection capabilities can be improved by including I_{DDQ} measurements [Naik93]. I_{DDQ} measurement is based on the fact that many defects result in an abnormal power supply current in the circuit. Measuring this current in the quiescent phase is called I_{DDQ} measurement [Maly88, Rodriguez91]. Furthermore, some of the non explained behaviour could be due to the basic assumption made on dealing only with catastrophic defects. It has been shown that non-catastrophic defects, especially in dense memory cells, result in delay faults [Walker92, Bruls93].

Validation of the Test Algorithm

The signatures used for the fault model validation analysis were also used for the test algorithm validation analysis. Each algorithm was assigned a *score* number. This number is the mean value taken over the signature bits from all devices of one test algorithm. Therefore, the maximum score that can be reached for a test algorithm is 2. The higher the score of a test algorithm, the higher its fault coverage of real faults compared to other test algorithms. The results of the test algorithm comparison study is shown in Table 4.2. The details of each test algorithm are indicated in the various rows of Table 4.2. For each test algorithm the following is given. A literature reference, an address order, the algorithm name, the number of data backgrounds used, an indication whether a data retention test has been used, and its resulting score number. The address order indicates the direction in which the memory cell array is traversed. This is either in an X address order, Y address order or both. The march test algorithms resulting from the Inductive Fault Analysis are called IFA_9N and IFA_13N. The best score was for the length 36N algorithm as proposed by Papachristou and Shagal running with four data backgrounds, twice with different address orders and a data retention test [Papachristou85]. Second and third are the IFA_13N and the IFA_9N algorithms.

4.1.6 Conclusions

Defect analysis using the Inductive Fault Analysis technique results into a physical failure based fault model. The proposed march test algorithms show excellent features in both test length and fault coverage. Both the fault model and march test algorithms have been validated in practice.

The march test algorithms cover 100% of the faults under the fault model and are independent of row, column and cell arrangement in the memory array. The choice of the test algorithm is slightly design dependent. A length 9N march test algorithm is given for SRAMs with combinatorial read/write logic. A length 13N march test algorithm is given for SRAMs with sequential read/write logic. The given march test algorithms are both suitable for bit and word oriented SRAMs.

It is important to notice that roughly 50% - 60% of the defects behave as stuck-at faults, i.e., memory stuck-at faults and total device failures. In the theoretical approach, single defects were assumed and 50% stuck-at was estimated.

90% of the analyzed defects could be explained by using the Inductive Fault Analysis technique and are covered by the fault model. Almost all fault classes as indicated in the fault model indeed occurred regularly in actual devices. The 10% unexplained defects have been studied in further research activities on I_{DDQ} measuring. Via I_{DDQ} measurement a further improvement in defect modeling and fault modeling techniques can be achieved. A next step in this improvement process is to take a closer look at non-catastrophic defects and corresponding delay faults.

The results of the Inductive Fault Analysis technique are dependent on the actual layout of the device and the manufacturing technology used. One cannot take over the results from the SRAM Inductive Fault Analysis for other memories without careful consideration. To circumvent the risk to repeat the Inductive Fault Analysis study too often, we argue that application of the Inductive Fault Analysis technique requires a stable design library where results can be used for a relatively long period.

Literature reference	address order	algorithm name	# data-back-grounds	data retention test	score
	Y	PRODPAT	1	no	0.81
[Nair79]	X	MATSPLS	1	no	0.14
[Nair79]	Y	MATSPLS	1	yes	0.37
[Nair79]	JUMP	MATSPLS	4	no	1.06
[Marinescu82]	JUMP	MARIN_C	1	yes	1.51
[Marinescu82]	X	MARIN_C	4	no	1.22
[Marinescu82]	Y	MARIN_C	4	yes	1.78
[Marinescu82]	X	MARIN_B	1	yes	1.64
[Marinescu82]	X+Y	MARIN_B	4	no	1.67
[Abadir83]	X	MARCH_01	1	yes	0.69
[Abadir83]	Y	MARCH_01	4	no	1.57
[Suk81]	Y	SUCK-RD	4	no	1.56
[Suk81]	X+Y	SUCK-RD	1	yes	1.76
[Nair78]	X	NAIR.AL	4	yes	1.68
[Nair78]	X+Y	NAIR.AL	1	no	1.64
[Papachristou85]	Y	PAPACHR	1	no	1.48
[Papachristou85]	X+Y	PAPACHR	4	yes	1.91
	Y	IFA_9N	4	yes	1.81
	Y	IFA_13N	4	yes	1.83
	Y	IFA_13N	4	no	1.61
	Y	FASTCHK	4	no	0.12
[Hayes75]		PATSENS		no	0.78

Table 4.2. The score of the test algorithms.

4.2 Built-In Self-Test for Static Random Access Memories

4.2.1 Introduction and Motivation

The size of embedded SRAMs as used in currently available devices is growing. A size of 2k words of 8 bits wide is common practice. When using the 9N or 13N SRAM march test algorithm, with possibly multiple data backgrounds, the total test data volume for this size of SRAM is at minimum 9 • 2k • #data-backgrounds per SRAM port. When this memory is made accessible via scan, the total test data volume has to be multiplied by the length of the scan chain. This leads to a high test data volume required on the IC tester. The vast majority of the total device test time is determined by the embedded memory test data volume. To limit this test data volume, alternative access protocols are required which results in a limited tester data volume.

A solution to this problem is providing a self-test capability to the embedded memory. The self-test executes the test algorithm and the only data required on the IC tester is the initialization data and the final response data. By offering a self-test capability, a design team has a choice what to do. In case of a performance-critical and die-critical design, the choice could be to have a longer test time. In case of tester data volume limitations, the choice could be for a self-test version. Offering this flexibility is of importance for optimizing towards a certain design or test constraint; see Chapter 3.

In this section we describe, as an example, the implementation of the 13N algorithm in a self-test machine for Static Random Access Memories [Dekker87, Dekker89]. Both the design specification and the design implementation is outlined.

4.2.2 Specification and Architecture of the Self-Test Machine

Specifications

The user requirements specification for the self-test machine can be stated as follows.

1. Embedded memories are considered as a yield determining part of the device. Therefore the test algorithm must have excellent fault detection capabilities. Hence, the test algorithm must be based on a realistic fault model and cover a large percentage of the faults.
2. To minimize the test costs, the test algorithm must be as efficient as possible.
3. Self-test limits the diagnostic possibilities. Without any precaution, it is difficult to find out which test pattern detected a defect. Defect location requires an additional serial scan access to all SRAM registers via which additional test patterns can be applied and observed.
4. It must be possible to apply a decent test to the self-test logic.
5. The influence of the additional self-test logic on the performance of the device must be minimal.
6. The amount of silicon real estate used for the self-test hardware must be minimal.

The Test Algorithm

As an example of an SRAM Built-In Self-Test implementation, the 13N march test algorithm is implemented, plus a data retention test. This march test algorithm is based on a realistic fault model, including stuck-at faults, stuck-open faults, transition faults, coupling faults and data retention faults. The test algorithm is discussed in detail in the previous section of this chapter.

The 13N march test algorithm is given in Figure 4.11. The required wait-time for the data retention test largely depends on the design. For the self-test implementation a wait-time of 100msec is taken. To detect coupling faults between different bit positions in one word of a memory cell, the minimal set of data backgrounds as derived in the previous section is used.

A self-test protocol for a word oriented SRAM is as follows.

* Initialize the self-test registers.
* Run the 13N march test algorithm consecutively for each data background.
* Run the data retention test with the last data background.
* Observe the self-test response.

Global Design Requirements

There are several global design requirements for implementing a self-test machine for a static RAM, which can be summarized as follows.

1. A synchronous operation of the SRAM is required for executing the self-test.
2. A provision for initialization of the self-test machine in a predetermined state is required.
3. It must be possible to isolate the SRAM during self-test from its environment. During self-test, the SRAM contents may not be disturbed by any signal from its environment. The only data it may receive is the test data required by the 13N march test algorithm.
4. A provision for observing the result of the self-test is required.
5. A provision is required for controlling the initialization, start, stop and observation of the self-test including controlling the execution of the march test algorithm.

It is assumed that all inputs of the SRAM, address, data, and control, are preceded by register cells. This solves the design requirement for a synchronous operation of the SRAM.

Initialization of the self-test logic can be done in various ways, for example by a reset signal or via serial scan access of initialization data. To circumvent the limited defect location capability of Built-In Self-Test and to ease the way of initialization it is an obvious choice to provide an additional scan functionality to all the register cells. This additional scan functionality provides the possibility to further test the SRAM for detailed defect information and to test the interconnect between the SRAM and its environment.

Isolation of the SRAM during self-test is obtained by providing the registers with a third mode of operation, called self-test. During this mode of operation the registers receive their data from either the self-test logic or the SRAM, but not from the environment.

The result of the self-test is compared with the known-good response. This can either be done by comparing the self-test response data with the known-good output or by compacting the self-test response data in a signature register. For several reasons the choice is made for a multiple input signature register for data compaction. During the last clock cycle of the self-test this signature register stores the final result which has to be observed and compared with the precomputed known-good signature. To guarantee correct observation of the final self-test signature the signature register is applied with a hold-mode. Upon finishing of the self-test, the signature register keeps its internal data valid until the moment it is scanned out for observation.

In total two test control signals, called ST_scan and ST_run, are necessary for controlling the registers cells. The signal ST_scan controls the register cells and enables the operation mode of normal or scan. The third register mode for self-test is controlled by the signal ST_run. Figure 4.12 presents a possible implementation of the register cells. Whenever needed, small modifications have been made on this basic design. The modifications are indicated in the figures when necessary.

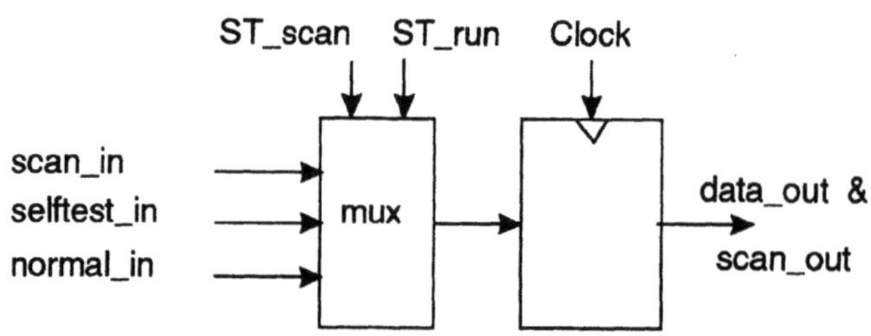

Figure 4.12. Implementation of the basic register cell.

Global Architecture

As can be derived from the 13N march test algorithm the following logical blocks are necessary.

1. An *address generator* that generates the upwards and downwards address sequences 0...N-1 and N-1...0.
2. A *wait-counter* is required to count down the data retention wait-time. This counter can be combined with the address generator.
3. A *data generator* that generates the data backgrounds and inverted data backgrounds.
4. A *response analyzer* that stores or compares the data received from the SRAM during a read operation.
5. A *self-test controller* that implements the 13N march test algorithm. This includes control of the SRAM and the other blocks of self-test logic. During the self-test the self-test controller has full control over the actions in the self-test machine.

The global architecture for a self-testable SRAM is given in Figure 4.13.

In the following section the various blocks in the self-test machine are outlined in more detail. Full details can be found in [Dekker87].

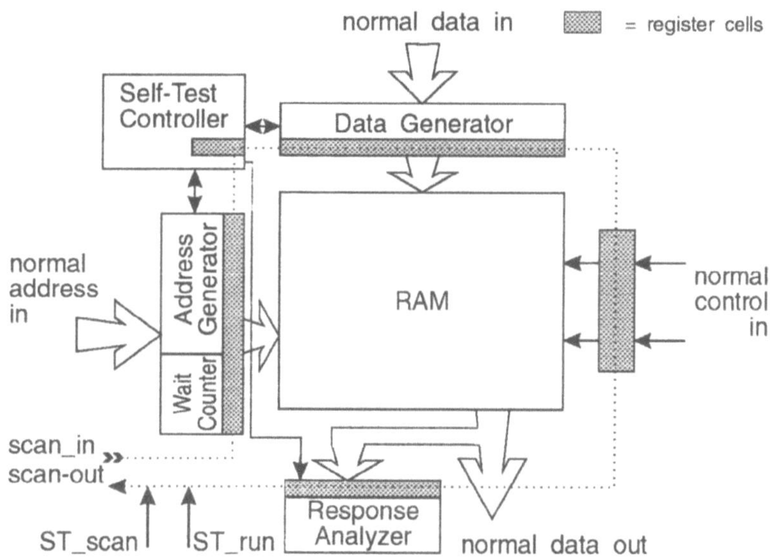

Figure 4.13. The architecture of a self-testable SRAM.

4.2.3 The Various Blocks of the Self-Test Machine

The Self-Test Controller

From the test algorithm with data backgrounds and data retention test a specification of the self-test controller is obtained in the form of a state diagram; see Figure 4.14. The regular structure of the test algorithm resulted in a state diagram specification of low complexity and in a low self-test controller complexity. Figure 4.15 gives a possible circuit realization of the self-test controller. The Programmable Logic Array (PLA) can be replaced by other combinational logic with the same functionality, e.g., a standard cell realization.

The PLA contains all information about the test algorithm. Hence, if another test algorithm with the same addressing sequence is used, only the PLA needs to be redesigned. The most essential input signals for the self-test controller are whether the last address has been generated (last address), the last data background has been applied (last dbgr), or whether the wait-time has been expired (end of wait).

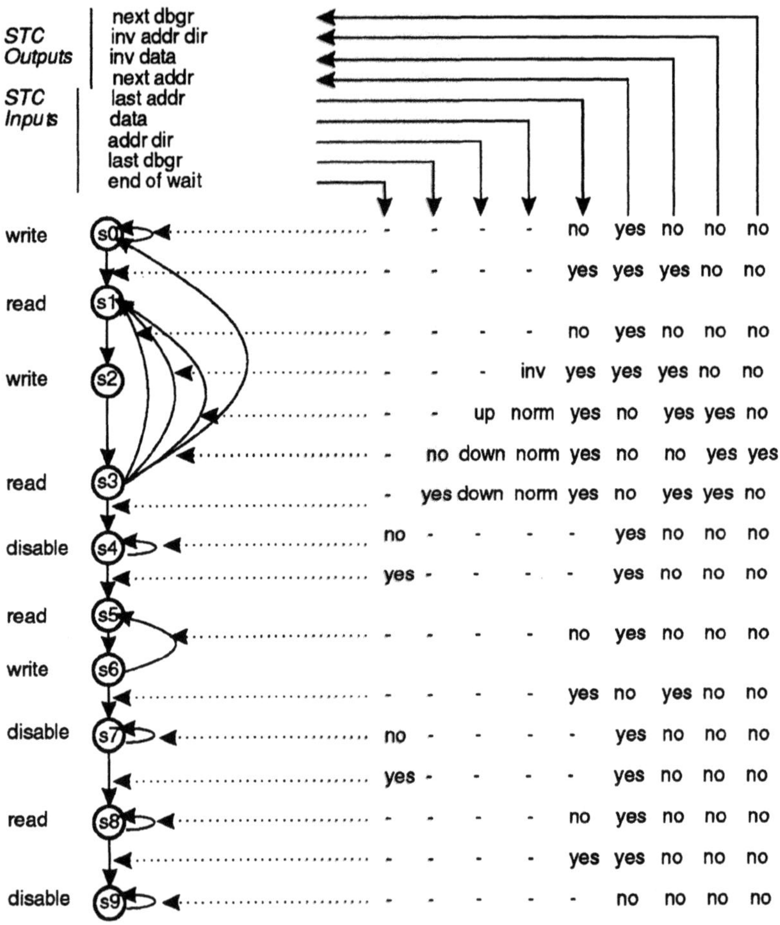

Figure 4.14. Self-test controller state diagram.

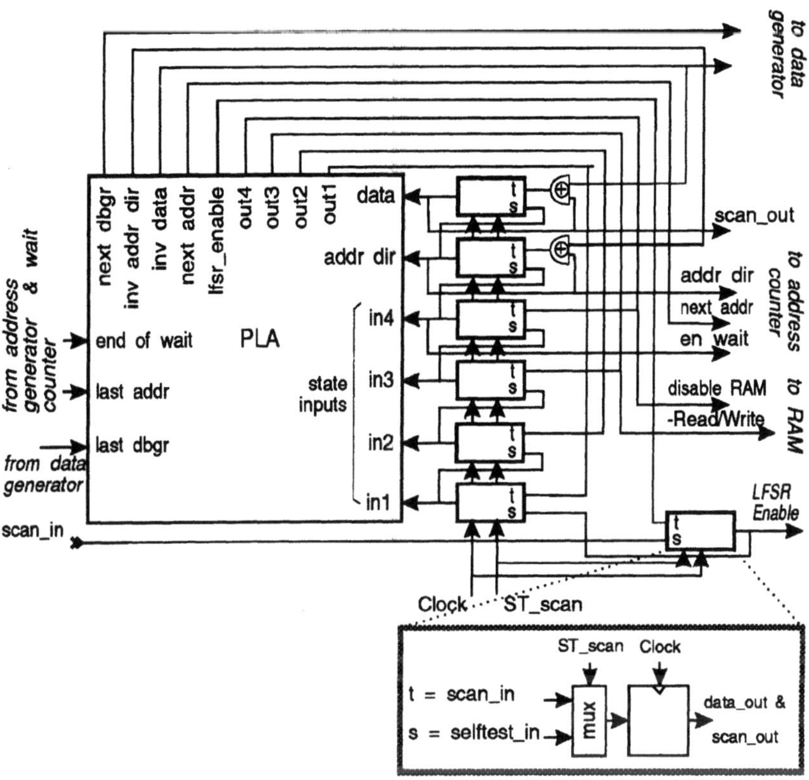

Figure 4.15. Circuit realization of the self-test controller.

The Data Generator

In self-test mode, the data register of the SRAM has the function of a data generator. The data that has to be generated during the test are the data backgrounds and the inverted data backgrounds. As an example the design of a data generator generating the data backgrounds as depicted in Figure 4.16 is given.

1. 01010101	2. 00110011
3. 00001111	4. 00000000

Figure 4.16. Four data backgrounds for a byte oriented SRAM.

There is a simple relationship between the bits of the data backgrounds of Figure 4.16. If a next data background is generated, bit 1 and 2 receive the value of bit 1 in the previous data background, bit 3 and bit 4 receive the value of bit 2 in the previous data background, etc. Due to this simple relationship the data generator circuit realization can be regular also. A possible circuit realization of a data generator for a byte oriented SRAM is given in Figure 4.17.

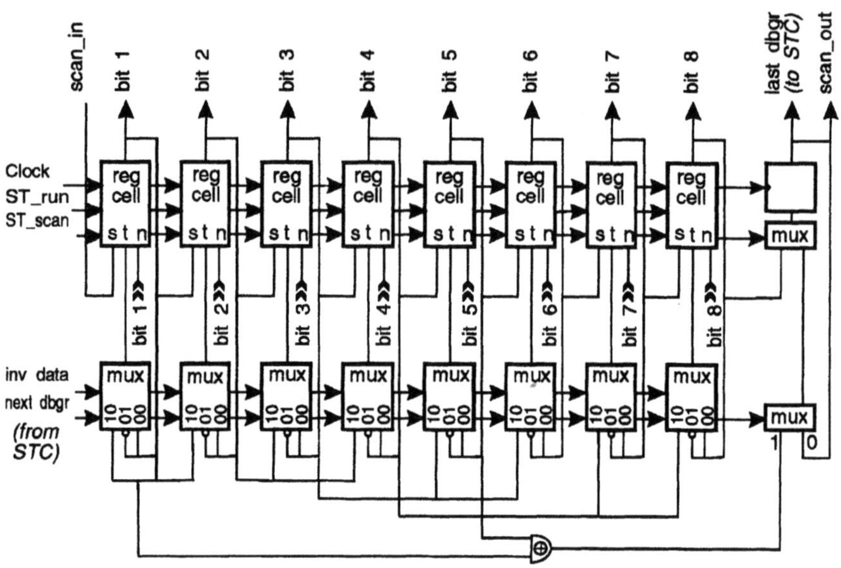

Figure 4.17. Circuit realization of the data generator.

The data generator is controlled by two signals from the self-test controller indicating 'next data background' (next dbgr) and 'invert present data background' (inv data). The register cells in the data generator are as given in Figure 4.12 and are controlled by the two signals ST_scan and ST_run. The data register is initialized with the first data background via scan. A stop criteria for the data generator is the indication of the last data background in present use (last dbgr). Knowing the order and characteristics of the data backgrounds to be generated, this detection is done via a simple exor-gate.

The data generator is linear parameterizable with respect to the width of the data backgrounds and it is compact to implement around an existing data register or register file.

The Address Generator and the Wait-Counter

The address generator and wait-counter are combined in order to reduce the use of silicon real estate. This is possible since no address is needed if the count down of the wait-time is active. The number of address bits plus the number of wait-counter bits, in combination with the clock frequency define the total wait-time for the data retention test. A stop criteria for the address generator is an indication of 'last address present' (last addr).

The address generator needs to have the following functional properties.

* Generation of all possible $N = 2^n$ addresses, in an arbitrary order, where N is the number of addresses and n is the number of address bits.
* Both an upwards and a downwards addressing sequence is required. Hence, it must be possible to reverse the address order.
* During a march element in the march test algorithm the address remains unchanged. It must therefore be possible to stop the address generator from generating new addresses.
* An indication of 'last address present' has to be generated since this determines a stop criteria.
* The address generator should be designed parameterizable with respect to the number n of address bits.

Two possible circuit realizations of the address generator have been taken into account: a linear feedback shift register (LFSR) and a binary up/down counter. A binary up/down counter has been chosen since this resulted in a smaller circuit. For example, the generation of the all-zero address via an LFSR requires additional hardware [Wang86].

A problem with the binary counter is the delay of the ripple carry signal. Including the wait-counter the number of stages in the counter can be up to 20 bits. This problem has been solved by designing an up/down counter with a pipelined carry signal. The maximum delay was thus reduced to the delay of a series of four 2-input AND gates.

Figure 4.18 shows an example of the circuit realization of a combined address generator/wait-counter.

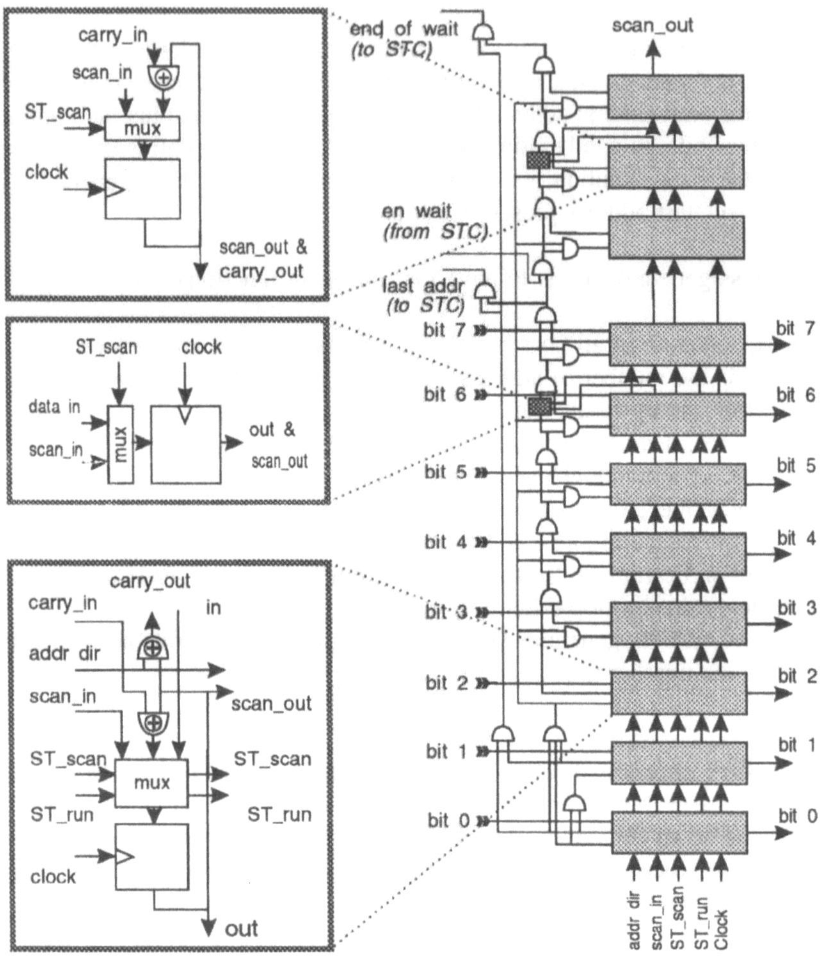

Figure 4.18. Circuit realization of the address generator/wait-counter.

The Response Analyzer

The data that is produced by the RAM during the read operations of the test algorithm, is received by the response analyzer. We distinguish the following two ways to analyze the response data with the expected data.

1. Compare during every read operation the data read with the expected data and set a pass/fail bit if a difference is detected.

2. Compaction of the data by polynomial division in a multiple input linear feedback shift register (MISR) The final contents of the MISR is called a signature and is compared with a pre-computed response.

The MISR implementation has been chosen since this approach results in the least silicon overhead and the delay of a MISR is small and independent of the number of bits.

After the self-test has been finished, the final signature needs to be compared with a signature that is previously calculated by a dedicated software tool. This comparison can be done on or off chip. On chip comparison results in one signal indicating a pass or fail of the self-test. Off chip comparison requires a comparison of all bits of the signature with their expected value. In order to circumvent the risk of a stuck-at fault in the pass/fail signal we have chosen to perform the comparison off chip. Therefore, the signature needs to be shifted out of the multiple input signature register after the self-test has been finished. A hold mode on the MISR is used in order to prevent that the final signature in the MISR is lost after the self-test has been finished. The hold mode control signal (LFSR enable) is generated in the self-test controller.

The aliasing probability of the MISR reduces exponentially with the number of MISR stages [Smith80, Bhavsar81, Carter82, Bardell87]. A minimum of eight MISR stages is considered a requirement. Normally, the number of MISR stages equals the number of data outputs of the SRAM. If the number of data outputs of the SRAM is smaller than eight, the remaining MISR inputs are set to constant 0 or 1.

Figure 4.19 shows a possible circuit realization of an eight bit MISR. A lot of research has been published on which feedback polynomial to use in the MISR to ensure a maximum error coverage and to minimize the aliasing probability [Bardell87, Agrawal93a, Agrawal93b]. It is commonly accepted that any primitive polynomial used in the MISR does provide acceptable results. The choice of the initial value of the MISR is arbitrary. The all-zero word has been taken for convenience.

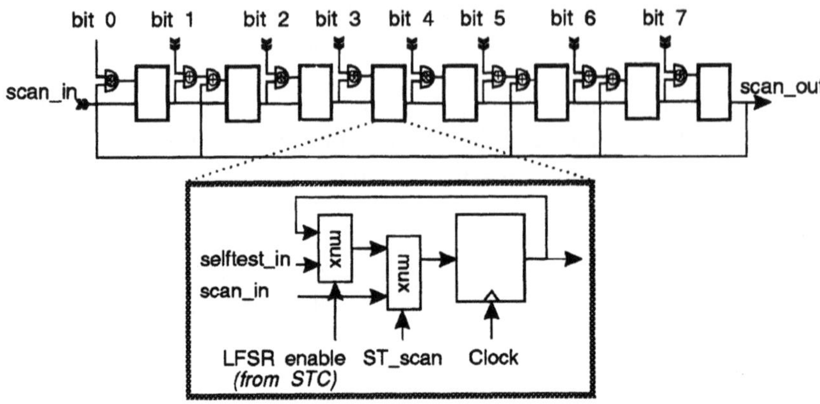

Figure 4.19. Circuit realization of an eight bit response analyzer.

4.3 Leaf-Macro Testability Study Aspects

In this chapter, examples have been given of leaf-macro testability studies. The presented approaches have been different in concept. The first example focused on test pattern generation based on an Inductive Fault Analysis approach. The second example discussed how to provide a Built-In Self-Test option with a leaf-macro module compiler. What are items to be considered during a study on leaf-macro testability and what conclusions can be drawn from the examples in this chapter?

There are several aspects which play a role in a leaf-macro testability study. The aspects we have encountered are discussed shortly.

Test Objectives

The test objectives, as globally stated in Chapter 1, have to be defined for each target leaf-macro. These test objectives define the effort which has to take place on defect modeling and the test pattern defect detection/defect location capabilities. It would be ideal if every leaf-macro is treated by an Inductive Fault Analysis approach. As is shown in Chapter 2, this is not always possible. What can be done more easily is a 'defect simulation' of a given set of test patterns.

Defect simulation refers to transistor-level circuit simulation with a *given* test pattern set as input and with a sample of manufacturing defects modeled in the transistor network. In this manner, weak spots in the set of test patterns can be detected and removed.

A topic which should not be ignored is the set of test requirements when the device is mounted on a printed circuit board and, in a next stage, in a system rack. Field service requirements demand the possibility for easily accessible devices and for diagnostic capabilities. In this case, a hierarchical approach is required. Decisions on testability issues which are correct for device testing may be incorrect for system testing. For example, a choice for Built-In Self-Test may be discarded for device testing due an increasing device die size. On the opposite, system test aspects may highly benefit from Built-In Self-Test.

Boundary Conditions

The definition of testability, given in Chapter 1, takes boundary conditions into account. Boundary conditions normally refer to items such as number of pins required for test, silicon real estate to be used, performance influence, and test data volume. These boundary conditions reflect their requirements on the leaf-macro testability studies.

Test Pattern Generation Capabilities

In case of regular structures, such as multipliers and memories, test generation can possibly be done manually with assistance of software tools. However, in general, manual test pattern generation for large complex devices is not feasible. Software tools have to assist a test engineer in the process of creating test patterns. Hence, an implicit boundary condition is the availability of a set of software tools capable of meeting requirements on defect modeling, defect or fault coverage, and maximum number of test patterns. This holds for test patterns for every objective of the test program. Whether they are meant for structural testing, functional testing, or application-mode testing.

As such, design for testability is bounded by the capabilities of software tools. There is a balance between the design for testability approaches, e.g. scan design, and software tools, such as automatic test pattern generators for combinational circuits. This is one of the reasons that software tools emerge which take care of this balance. Examples of such tools are automatic scan cell inserters for full or partial scan and scan chain routers.

A special case is I_{DDQ} testing. I_{DDQ} testing requires that there is no power consumption during the quiescent period of the clock phase. Besides the difficulty of test pattern generation for I_{DDQ} faults, this puts requirements on the design style. Examples of such requirements are no use of active pull ups, no dynamic pass transistor logic, and control of the states of buses.

Access requirements

The test of a leaf-macro is applied from the device pinning. Access is created to and from every leaf-macro and various resources, such as pins and registers, of the device are used for this purpose. This has several consequences.

First of all, a leaf-macro test method requiring a complex application protocol might give severe design requirements for accessing this leaf-macro. An example is a leaf-macro with a partial scan approach. A range of test patterns is required to detect a fault in a sequential network. These test patterns have to be applied in a predetermined order. Each fault may result in a different number of required test patterns. This means that the application protocol is fault specific. Consequently, access to such a leaf-macro almost certainly demands a direct access from the device pins to the leaf-macro ports.

Secondly, the choice of access means for a leaf-macro influences the possibilities for an optimization of the final device program during the test assembly phase. This optimization might become complicated or even impossible. As will be shown in Chapter 7, device resources are used for leaf-macro access creation. Leaf-macro test specifications can be assembled into a device test specification in a test data volume minimal way when there are no resource conflicts. Resources are used either for data or for test control. The less the amount of non-conflicting resources and the more compatible the various test control sequences, the more the final device test specification can be optimized for minimal test data volume.

Finally, system test requirements may impose access to leaf-macros from the system level ports. Extending Macro Test to system level itself is possible. However, in this case, a correct approach is to apply Built-In Self-Test techniques to leaf-macros. Built-In Self-Test requires a minimal amount of resources to access the corresponding leaf-macro and the test is locally executed. It therefore provides facilities for hierarchical testing.

From these various aspects the following conclusions can be drawn.

* It is preferable to apply Inductive Fault Analysis techniques for creating and verifying test patterns. This is certainly feasible in case of relatively small leaf-macros or leaf-macros consisting of an array of identical elements, such as ROMs, RAMs, PLAs, array multipliers.

* The theory and application of I_{DDQ} measurement is hardly treated in this book. Research has shown that I_{DDQ} measurement is a good technique to be used in conjunction with other structural test techniques such as Inductive Fault Analysis [Maly89, Soden89, Maxwell92, Naik93]. It is therefore recommended that the design of a leaf-macro should not circumvent an I_{DDQ} measurement.

* The leaf-macro test method should not complicate the access protocol from the device pins unnecessarily. A scan testable leaf-macro provides an easy access protocol and is therefore a preferred approach.

* Make use of Built-In Self-Test techniques when there is a limit on total test data volume and when requirements exist for system level testing.

* Provide a test method for every leaf-macro residing in a design library. In this manner, the test is constructed only once and used by many others. A prerequisite is that the design library is stable and does not change too often.

5 Scan Chain Routing with Minimal Test Application Time

5.1 Leaf-Macro Access

Macro Test requires access to every leaf-macro in the design. Access is required for both the data flow and the control signals enabling the data flow. In general, choices can be made in establishing this access. The choices can be grouped into two categories. Either, use is made of functionally existing paths and control signals, or design entities are added to create the required access. The latter has to be used when the first possibility is not available. Which choice to take depends mainly on the complexity of the initial test plan, the design architecture, the availability of transfer information, and the constraints on number of pins, use of silicon area, performance, and test data volume; see Chapter 3. The set of design entities and device pins involved in creating the access for leaf-macro test data and control is called the *leaf-macro access path*. The influence of the choice of the initial test plan can best be seen when comparing a leaf-macro tested with Built-In Self-Test or with partial scan. A Built-In Self-Test function only needs to be initialized, executed, and observed. Only a few leaf-macro terminals have to be accessible; see Chapter 4. On the other hand, for a partial scannable leaf-macro, all terminals have to be accessible with a complicated timing protocol added to it. In this comparison, the price of an easy access is a set of local design adaptions for data and control generation for the Built-In Self-Test function.

An obvious requirement for access is that the leaf-macro access path does not hamper the normal functional operation of a design. It is because of this reason that the data paths and control signals resulting from additional design entities are kept separate from functional data paths and control signals. The control signals of these additional design elements are therefore called *test control* signals. Examples are the scan-enable signal of a scan cell and the initialize/observe and execute signals of Built-In Self-Test logic.

When adding design entities for access creation, people often make use of standard access mechanisms. The Boundary Scan standard is an excellent example of this [IEEE90]. Another standard access method is scan design. Standards are easy to interpret and easy to implement automatically. The disadvantage of standard access mechanisms is that they are standard. This means that in cases where better solutions are available because of e.g., a certain design architecture, use of the standard access method may require too much silicon real estate, influence the performance too much, and result in an excessively long test time. For this reason, effort is spent on methods that almost meet the standard access and minimize on the penalties. An example is the partial scan technique [Agrawal87, Ma88, Lee90]. A comparison between full scan and partial scan can be found in [Bennetts93].

With Macro Test, the possibility exists to create access to each leaf-macro in a design specific way. Use can be made of specific design features, for example, buses and transfer information. Every decision influences the use of silicon real estate, test data volume, pinning, etc. We have chosen for a set-up with a separation between testability synthesis and test synthesis activities. This requires that especially decisions which influence the final test data volume have to be made visible, and therefore become manageable, during the testability synthesis phase. During the test plan generation process, the decisions for the access paths are made. When assembling the various leaf-macro test specifications into a device test specification, the consequence of the choices become visible. There is an associated resource allocation problem with this. It might be that a certain choice for an access path to a certain leaf-macro influences the test data volume. An easy example is the choice of placing the scannable inputs of an SRAM at the end of a long scan chain. The shifting process of a large amount of input test patterns results in an excessively long test time.

To model and handle problems, we have separated a number of activities for creating leaf-macro access. First, a distinction is made between test data flow and test control flow. Second, optimization techniques are considered which reduce the test time by exploiting parallelism in leaf-macro access. These activities are the topic of the following chapters.

Chapter 5 deals with the topic of *scan chain routing*. Most of the applied access mechanisms rely on the use of serial scan access. One of the scan access penalties is the increase in test time due to the shifting process. This penalty is minimized by carefully connecting the various scan registers in the design. In Chapter 6, the *test control signals* are considered. As stated, these signals are kept separate from the other functional signals. When no separate action is taken they have to be connected to separate device pins which increases the pin count. The penalty of pin count increase is minimized by the design of an *on-chip controller* which uses a minimum number of device pins. In Chapter 7, a general theory is given on test time minimization by exploiting parallelism in testing leaf-macros. Parallelism can be exploited when there is a minimal resource sharing between the various generated test plans. To circumvent certain resource conflicts it might be that alternative access choices for leaf-macros have to be made. This creates a loop in the testability synthesis activities. For example, when access is initially provided via internal buses, only one leaf-macro is allowed to write onto the bus. Furthermore, the bus protocol has to be obeyed at any time. When this circumvents the parallel execution of tests for various leaf-macros, then the choice could be made of not using buses for access.

5.2 Introduction to Scan Chain Routing

A widely applied technique to generate access to internal nodes of a device is *scan design* [Williams73, Eichelbrg78, Williams83, Bennetts84, McCluskey84, Fujiwara85]. In the scan design technique, each of the circuit flip-flops is replaced by a flip-flop structure where a multiplexer is placed at the data input to permit a selection of two different data inputs. Such a flip-flop is called a *scan cell*. One input is considered during normal operation and the other input is considered during test mode. If a device is in test mode, all flip-flops in the circuit are chained to form one or more shift register structures, called *scan chains*. All the states of the scan cells can be controlled and observed serially from the device input pins and output pins; see Figure 5.1. Besides increased accessibility, the scan design technique has the important benefit that sequential circuit test pattern generation is not required. Test pattern generation need only be done for combinational circuits since the flip-flops can be accessed and tested directly.

A drawback of scan design is the time needed for serially shifting all data. Large designs may have thousands of flip-flops, resulting in excessive test times. Especially, when the design has embedded memories, where an abundance of data is involved, this drawback becomes severe. Therefore, an approach is required to reduce this penalty to a minimum by carefully routing the scan cells and scan registers into scan chains. A *scan register* is a set of already routed scan cells and treated as one design entity. A large number of possible scan chain routings exists; see Figure 5.2. Each routing has an associated cost in terms of test time. The problem is to find the routing having the minimal cost.

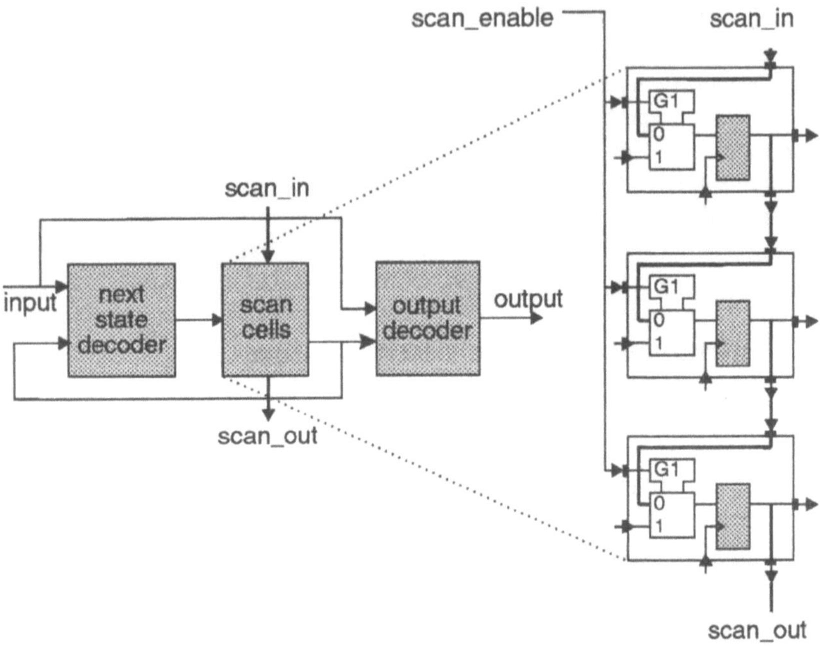

Figure 5.1. Principles of scan design.

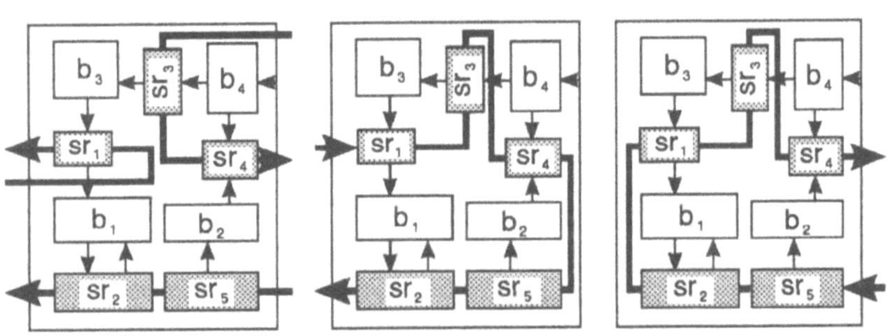

Figure 5.2. Some possible scan chain routings in an example circuit.

The basic idea of test time optimal scan chain routing is easy to explain. When we have a leaf-macro with a large set of test patterns we place the *pattern-applying registers* for this leaf-macro close to the scan data input and the *pattern-observing registers* close to the scan data output. This informal idea is formalized in a mathematical theory. It is shown that finding the optimal solution is an NP-complete problem.

5.3 Scan Test Application Protocol

Before defining the scan chain routing problem, several assumptions have to be made. We deal with the routing of scan registers into scan chains. Scan registers are considered as resources for applying and observing test data. Optimal use of resources leads to an optimal device test specification in terms of volume of test data and total test application time. Optimal resource allocation is in itself a complex optimization problem. The problem of scan chain routing is tightly coupled to the problem of optimal resource allocation. To be able to handle this interaction we have decoupled the two items. We first treat the topic of scan chain routing and in Chapter 7 we treat the topic of test optimization by optimal resource usage.

Due to this assumption we consider the following test application protocol for a scan-testable device.

1. Scan in input test data.
2. Apply and observe the test data.
3. Scan out the response test data.

Step 1 and 3 are done serially. Step 2 is done in parallel.

As such, the simultaneous action of shifting in new test data while shifting out the response data is not taken into account. We consider this action as a possible step towards optimal resource allocation, i.e., making use of the same resource, however at a different timing cycle.

5.4 Scan Chain Routing Problem Formulation

Up to now, we have regarded a circuit to consist of leaf-macros. For the scan chain routing problem we take another look at a circuit. Since scan registers are the main entities we will deal with, we consider a circuit to be build up out of 'scan registers and the rest'. By extracting the scan registers from the circuit, and hence also from the contents of a leaf-macro, we may regard the circuit to consist of a set of scan registers and a set of logic blocks. Each scan register is acting as a pattern-applying, pattern-observing, or both, register for one or more blocks. It is primarily the set of scan registers and the test data volume per scan register we are interested in. We formalize this in the *Scan Chain Routing Problem.* The problem statement is based on the following assumptions.

* All scan registers are operated synchronously.
* Scan registers may be allocated freely to scan chains.
* Scan registers are considered to be predefined design entities. This means that they cannot be split into multiple pieces.
* The scan test application protocol is equal for each scan vector and consists of a number of scan cycles for shifting in the stimuli data, one cycle for applying/observing the data followed by a number of shift cycles for observing the serial response data.
* Scan registers are scanned in and scanned out completely. Shifting data into and out of the scan register requires a number of shift cycles equal to the scan register length.

Note 1. Certain defects may result in a sequential behaviour. Testing for these kind of defects may require a two-pattern test. One initialization pattern and one detection pattern. Application of these two-pattern tests via scan is not always possible. The shifting process of the second pattern may disturb the initialization of the first pattern. A solution to this problem is to modify the scan cell or to modify the order of the test patterns. The test patterns can be modified by taking care that pattern 2 is overlaid with pattern 1 such that it can be applied after 1 additional shift cycle. Examples of such techniques are described in [Leenstra93b]. Such an overlay approach is not considered in this chapter.

Scan Chain Routing Problem SCRP

Given is a circuit **B** that is partitioned into m > 0 logic blocks according to,

$$\mathbf{B} = \{b_1, b_2, ..., b_m\},$$

and a set of n > 0 scan registers **SR**,

$$\mathbf{SR} = \{sr_1, sr_2, ..., sr_n\}.$$

Given is also a set **T** of s > 0 tests to be applied and observed to the circuit via the scan registers

$$\mathbf{T} = \{t_1, t_2, ..., t_s\}.$$

A test t_i is a *set* of leaf-macro test patterns and s equals the number of leaf-macros.

Scan registers may be routed into one or more scan chains. Each possible routing results in a *scan chain configuration.* The set **SCC** of all possible scan chain configurations is given by

$$\mathbf{SCC} = \{scc_1, scc_2, ..., scc_M\}, M > 0.$$

For each scc_i, $1 \leq i \leq M$, the number **k**: $scc_i \rightarrow \mathbf{N}$ indicates the number of scan chains involved, and a cost **C**: $scc_i \rightarrow \mathbf{N}$ is proportional to the number of shift cycles for applying and observing all tests in **T**. **N** is the set of the natural numbers.

The Scan Chain Routing Problem SCRP is defined as the problem

$$\text{find } scc^* \in \mathbf{SCC} \text{ such that } \mathbf{C}(scc^*) \leq \mathbf{C}(scc_i) \text{ for all } scc_i \in \mathbf{SCC}. \quad (5.1)$$

In other words, the Scan Chain Routing Problem is to minimize the overall test time of the circuit by configuring the scan registers sr_i in one or more scan chains such that the number of serial shifting cycles of test **T**, for applying stimuli and observing responses to and from blocks in **B**, is minimal.

We define the cost $C(scc_i)$ more precisely and deduce some properties of the search space involved in this scan chain routing problem.

In practice, there are boundary conditions stated by the design team on maximum number of scan chains and maximum cost. It may be possible that the targets cannot be met. In this case there are two alternative decisions which can be made. Either relax the boundary conditions or search for another solution not based on serial scan access resulting in less test time.

5.5 Scan Chain Routing Cost Model

The model of the circuit and its tests consists of four components. Three circuit components being scan registers, scan chains, and scan chain configurations, and one test data component being the set **T**. The cost function, associated with a certain scan chain configuration, must reflect the number of shifting cycles involved in applying all tests in the set **T**. The cost model contains contributions of these four components.

Scan Registers, Scan Chains and Scan Chain Configuration

A circuit is divided into a set of logic blocks **B** and a set of scan registers **SR**. A scan register $sr \in$ **SR** consists of a number of consecutive scannable flip-flops and forms one single design entity. All available scan registers are routed together to form one or more scan chains. Routing of scan registers means that the serial output port of one scan register is connected to the serial input port of another scan register or to an output pin of the device. The first serial input port of a scan chain is connected to an input pin of the device. As such, we may regard a scan chain as a *sequence* of scan registers. A given routing of scan registers into one or more scan chains results in a scan chain configuration. Hence, a scan register configuration is a *unordered set* of scan chains. If we denote a scan chain k by sc_k and scan chain configuration j by scc_j, then

$$\forall j,\ 1 \leq j \leq M,\ scc_j = \{sc_1, sc_2, \ldots, sc_{k(j)}\}$$

$$\forall k,\ 1 \leq k \leq k(j),\ sc_k = (sr_p, \ldots, sr_q), \text{ whereby}$$

$$sr_p \rightarrow sr_{p+1} \rightarrow \ldots \rightarrow sr_q.$$

The values p, ...,q are function of k.

The ordering relation $sr_x \rightarrow sr_y$ of two scan registers in the same scan chain means 'scan register sr_x is positioned before scan register sr_y, and sr_x and sr_y are not equal'.

An example of scan registers, a scan chain, and a scan chain configuration is shown in Figure 5.3.

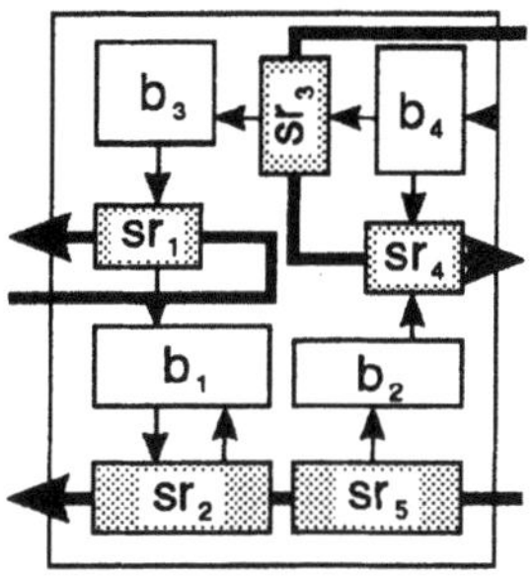

$SR = \{sr_1, sr_2, sr_3, sr_4, sr_5\}$

$sc_1 = (sr_5, sr_2)$
$sc_2 = (sr_1)$
$sc_3 = (sr_3, sr_4)$

$scc_1 = \{sc_1, sc_2, sc_3\}$

Figure 5.3. An example of scan registers, scan chains and a scan chain configuration.

A Test

With each test $t_i \in \mathbf{T}$ we associate three items: the number N of test patterns in t_i, the set of pattern-applying scan registers RA, and the set of pattern-observing scan registers RO.

$N(t_i) = | t_i |$,
$RA(t_i) = \{ sr \in \mathbf{SR} \mid sr \text{ is used to apply serially scanned data from } t_i\}$,
$RO(t_i) = \{ sr \in \mathbf{SR} \mid sr \text{ is used to observe results of the applied data from } t_i\}$.

The sets RA and RO are determined during test plan generation while searching for access to every leaf-macro. An example of RA and RO sets is shown in Figure 5.4. In this example, the circuit consists of four blocks and five scan registers. The test set **T** consists of three tests. Test t_1 applies data to b_1 and b_2 and observes data from b_3. Test t_2 applies data to b_3 and observes data from b_3. Test t_3 applies data to b_4 and observes data from b_2. From these given items, the applying and observing set for each test are easily found.

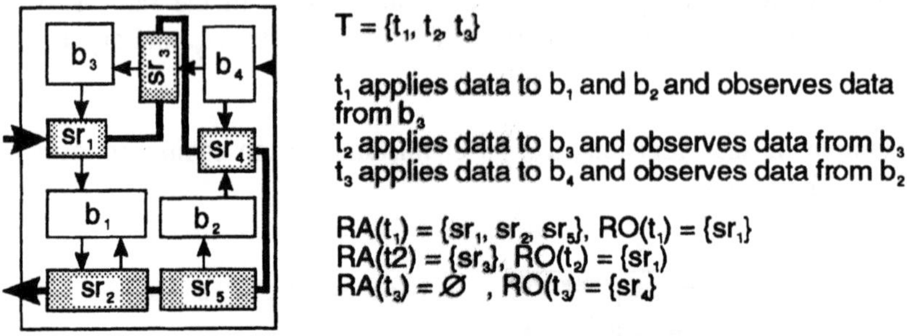

Figure 5.4. An example of tests with pattern-applying and pattern-observing scan registers.

Input and Output Distances

The number of shift cycles involved in applying and observing data via scan register sr in a given scan chain sc depends on the position of sr in sc. This is reflected in the following definition of *input distance* and *output distance*; see also Figure 5.5.

With a scan register sr within a scan chain sc of a given scan register configuration scc we associate the following number of parameters.

To apply stimulus data via scan register sr, we have to shift the patterns from the input of the scan chain sc to the last scan cell in scan register sr. This length is called the *input distance* $d^{in}(sr,sc)$. The same applies for shifting a response pattern from the first scan cell in sr to the scan output of the scan chain sc, which is called the *output distance* $d^{out}(sr,sc)$.

The number of scan cells in the scan register sr is the *scan register length* l(sr).

$$l(sr) \quad = \quad \text{\# scan cells in scan register sr.} \tag{5.2}$$

Hence, for a given scan chain sc = $\{ sr_p, ..., sr_i, ...,sr_q\}$, containing sr_i it holds that

$$d^{in}(sr_i,sc) = l_p + l_{p+1} + ... + l_i, \text{ and} \tag{5.3}$$

$$d^{out}(sr_i,sc) = l_i + l_{i+1} + ... + l_q, \tag{5.4}$$

where l_i is a shorthand notation for $l(sr_i)$.

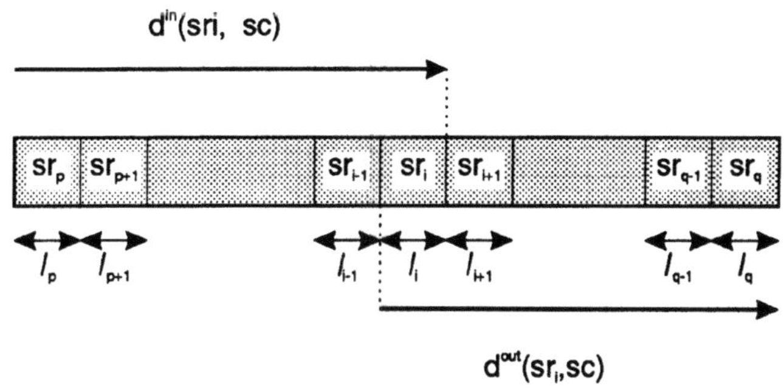

Figure 5.5. Input and output distances.

Scan Chain Cost

Having defined the important elements of our model, we can now define the cost $\check{C}$ associated with a certain scan chain sc, in a given scan chain configuration scc, and a test $t \in T$.

The cost $\check{C}(sc, t)$ of test $t \in T$ with scan chain sc is defined as the number of shift cycles necessary to shift in all stimulus patterns and to shift out all response patterns. These shift cycles are dependent on the input distances and output distances of the scan registers, involved in test t, which are element of scan chain sc. For this reason, we define the input distance d'^{in}, and output distance d'^{out} of a scan chain sc and a test t.

$$d'^{in}(sc, t) = \max \{d^{in}(sr_i, sc) \mid sr_i \in RA(t) \wedge sr_i \text{ in } sc\}, \tag{5.5}$$

$$d'^{out}(sc, t) = \max \{d^{out}(sr_i, sc) \mid sr_i \in RO(t) \wedge sr_i \text{ in } sc\}. \tag{5.6}$$

We can now define the scan chain cost $\check{C}(sc, t)$ in terms of the input distance and output distance.

$$\check{C}(sc, t) \quad = \quad N(t) \cdot [\, d'^{in}(sc,t) + d'^{out}(sc, t) \,]. \tag{5.7}$$

Scan Chain Configuration Cost

Next, we assume that we can scan all scan chains of a given scan chain configuration in parallel. From this assumption, we can deduce from (5.7) the cost $\hat{C}(scc, t)$ of a scan chain configuration scc for test $t \in T$.

$$\hat{C}(scc, t) = \max \{ \check{C}(sc_i, t) \mid sc_i \in scc \}. \tag{5.8}$$

Now, we can finally derive the target cost function which has to be minimized and which is representing the number of scan cycles involved in applying all tests.

The total cost C(scc) for a given scan chain configuration scc is defined as the summed cost of all separate costs $\hat{C}(scc, t)$.

$$C(scc) = \Sigma_{t \in T} \hat{C}(scc, t). \tag{5.9}$$

Note, that this cost function assumes that the scan-out cycle is not taking place concurrently with the scan-in cycle.

5.6 Scan Chain Routing Problem Complexity

Before treating the Scan Chain Routing Problem in more detail, we analyze the complexity of the problem itself. We explore the search space of this optimization problem and we prove that the decision variant of the Scan Chain Routing Problem is NP-complete.

The problem of scan chain routing is similar to the problem of partitioning a set of n elements into k subsets, whereby each subset has a quasi ordering relationship [Stanat77]. Based on the partition theory as described in [Hall67, Biggs85], we deduce some Scan Chain Routing Problem search space properties.

A scan chain is a sequence of scan registers, whereas a scan chain configuration is a set of scan chains. There is an ordering relation defined between the scan registers in the scan chain. The ordering relation $sr_i \rightarrow sr_j$ of two scan registers in the same scan chain means that scan register sr_i is positioned before scan register sr_j, and sr_i and sr_j are not equal. This ordering relation has the property that it is transitive, i.e., if $sr_x \rightarrow sr_y$ and $sr_y \rightarrow sr_z$, then $sr_x \rightarrow sr_z$. Furthermore, it is irreflexive, i.e., $\neg(sr \rightarrow sr)$ for all scan registers sr in a scan chain.

Theorem 5.1
Let SO(n,k) denote the number of possible scan chain configurations of n scan registers with k scan chains, where $1 \leq k \leq n$. Then

$$SO(n,1) = n! \quad (5.10)$$
$$SO(n,n) = 1 \quad (5.11)$$
$$SO(n,k) = SO(n-1,k-1) + (n+k-1)SO(n-1,k), \text{ for } 1 < k < n. \quad (5.12)$$

Proof
Consider the set of n scan registers, $\mathbf{SR} = \{sr_1, ..., sr_n\}$. There are n! ways to permute the elements of a set of n elements. Hence SO(n,1) = n!. Furthermore, there is only one scan chain configuration with n scan chains, the singleton sequences (sr_i), i = 1, ..., n.
Let us consider scan register $sr_n \in$ SR. A scan chain sc containing sr_n has the property that either (i) sc equals the singleton sequence (sr_n) or (ii) sc contains also other scan registers. When we remove the scan register sr_n from a scan chain configuration of type (i) we obtain a scan chain routing of (n-1) scan registers of the set $\mathbf{SR}\backslash\{sr_n\}$ into (k-1) scan chains. There are SO(n-1,k-1) of such scan chain configurations. Conversely, if we are given such a scan chain configuration, we can restore the sequence (sr_n) to create a scan chain configuration of the set **SR** with k scan chains.
Now suppose we have a scan chain configuration P of type (ii) with scan chains $sc_1, sc_2, ..., sc_k$. There is exactly one scan chain sc_i which contains sr_n and some more scan registers. We now define the scan chain configuration P_o as follows. P_o is the scan chain configuration of the set $\mathbf{SR}\backslash\{sr_n\}$ with scan chains $sc_1, sc_2, ..., sc_i\backslash(sr_n), ..., sc_k$.

There are SO(n-1,k) possible scan chain configurations P_o of n-1 scan registers with k scan chains. Given a scan chain sc from such a scan chain configuration, we can add (sr_n) to sc in |sc|+1 manners, where |sc| denotes the number of scan registers in sc. Hence, in total we can add (sr_n) to P_o in the following number of ways.

$$\Sigma_{i=1..k} [\ |sc_i| + 1\ ] = \Sigma_{i=1..k} |sc_i| + k = (n-1) + k.$$

This completes the proof. □

As a consequence of Theorem 5.1, we can tabulate the SO(n,k) numbers in a triangle.

	k=1	k=2	k=3	k=5	k=5
n=1	1				
n=2	2	1			
n=3	6	6	1		
n=4	24	36	12	1	
n=5	120	240	120	20	1

Table 5.1. SO(n,k) numbers in a triangle.

Note. The number of partitions of a set X of cardinality n into k disjoint subsets is denoted by S(n,k). The numbers S(n,k) are called the Stirling numbers of the second kind.

To get a better feeling for the order of magnitude of the SO(n,k) numbers we have deduced two identities.

Theorem 5.2

$$SO(n,2) = (n-1)n!/2 \qquad (5.13).$$

Proof

We prove this identity using the principle of induction.

For n = 2 it follows from (5.11) that SO(n,2) = 1. Hence, (5.13) is true for n = 2.

Suppose the theorem is true for n = N, that is,

$$SO(N,2) = (N-1)N!/2.$$

Then

$$\begin{aligned} SO(N+1,2) &= SO(N,1) + (N+1+2-1)SO(N,2) && \text{(by 5.12)} \\ &= N! + (N+2)SO(N,2) && \text{(by 5.10)} \\ &= N! + (N+2)(N-1)N!/2 && \text{(by induction hypothesis)} \\ &= N(N+1)!/2. \end{aligned}$$

So, the assumption for N implies the theorem for N+1. The theorem holds for n = 2 and thus by induction it holds for all n ≥ 2. □

Theorem 5.3

$$SO(n,n-1) = n(n-1). \tag{5.14}$$

Proof

For n = 2 it follows from (5.10) that SO(2,1)=2. Hence, (5.14) holds for n = 2. Suppose the theorem holds for n = N, that is,

$$SO(N,N-1) = N(N-1).$$

Then
$$\begin{aligned} SO(N+1,N) &= SO(N,N-1) + (N+1 + N -1)\cdot SO(N,N) && \text{(by 5.12)} \\ &= SO(N,N-1) + 2N\cdot SO(N,N) \\ &= N(N-1) + 2N\cdot SO(N,N) && \text{(by induction hypothesis)} \\ &= N(N-1) + 2N && \text{(by 5.11)} \\ &= (N+1)N. \end{aligned}$$

So, the assumption for N implies the theorem for N+1. The theorem holds for n = 2 and thus by induction it holds for all n ≥ 2. □

Note from Theorem 5.2 that the growth in size of the Scan Chain Routing Problem search space for only two scan chains is already in the order of magnitude of n!. The exponential size of potential search space is a necessary condition for a problem to be intractable and that the solution has to be found via heuristics and possibly exact solutions in some special cases. This indeed can be proven, which is the subject of the following Theorem.

First, we formulate the decision variant *SCRP* of the Scan Chain Routing Problem.

Let C_{max} denote a maximum cost. The Scan Chain Routing Problem *SCRP* is defined as the problem to

find a $scc^* \in SCC$ such that $C(scc^*) \leq C_{max}$.

Theorem 5.4

The Scan Chain Routing Problem *SCRP* is NP-complete.

Proof

The proof that the Scan Chain Routing Problem *SCRP* for a single scan chain is NP-complete can be found in [Narayanan92]. □

5.7 Routing of Scan Registers into a Single Scan Chain

Since the decision variant of the Scan Chain Routing Problem is NP-complete, we have to limit the search space or use heuristics to solve the problem in the case we have to deal with large problem instances. In this section, we take a closer look to the problem of routing scan registers into a single scan chain. Both mentioned approximation methods, limiting the search space and applying heuristics, are used.

The heuristics we use is the same as mentioned in section 5.2. Scan registers which are mainly used as a pattern-applying register are placed close to the scan chain input and scan registers which mainly used as a pattern-observing register are placed close to the scan chain output. A measure of 'usage' of a scan register is given based upon which a heuristic algorithm is given.

A way to limit the search space is to take only local criteria into account. The approach we take is to swap the positions of two *consecutive* scan registers in a scan chain if the cost function (5.9) decreases. A *local search* algorithm is presented which is based on this swapping principle and which uses an easily computable expression for the cost function difference.

Scan Register Properties

At the moment, the cost (5.9) is expressed as a sum over the tests t_i, with a maximum involved. For developing the heuristics and local search algorithms, we rather would like to have the cost expressed in terms of scan register properties. This gives the possibility to express the 'usage' of a scan register in terms of scan register properties and to relate it to the cost function. Furthermore, it gives the possibility to express a cost difference in terms of scan register properties when swapping the positions of two consecutive scan registers in a scan chain.

For each test $t \in \mathbf{T}$ we have an associated set RA(t) of applying scan registers and a set RO(t) of observing scan registers; see Section 5.5. Given these sets, we deduce for each scan register $sr \in \mathbf{SR}$ the number of times it is used for applying data and the number of times it is used to observe data. These numbers are given by the following formulas.

For $1 \leq i \leq n$

$$N^{in}(sr) = \Sigma_{t \in T} N(t) * Apply(sr,t), \tag{5.15}$$

$$N^{out}(sr) = \Sigma_{t \in T} N(t) * Observe(sr,t), \text{ where} \tag{5.16}$$

$$Apply(sr,t) = 1 \text{ if } sr \in RA(t), \quad = 0 \text{ else;}$$

$$Observe(sr,t) = 1 \text{ if } sr \in RO(t), \quad = 0 \text{ else.}$$

The maximum involved in the cost function (5.9) defines two sets of scan registers. These sets depend on the scan chain sc. For each $t \in \mathbf{T}$ there is a scan register $sr \in RA(t)$ the farthest from the scan chain input. Scanning of input data of test t requires scanning up to and including sr. Similarly, for each $t \in \mathbf{T}$ there is a scan register $sr \in RO(t)$ which is the farthest from the scan chain output. Scanning of response data of test t requires scanning from sr to the scan chain output. These scan registers define a set of farthest input scan registers $F^{in}(sc)$ and a set of farthest output scan registers $F^{out}(sc)$.

$$F^{in}(sc) = \cup_{t \in T} \{sr \mid sr \in RA(t), \forall_{sr^* \in RA(t)}[d^{in}(sr^*,sc) \leq d^{in}(sr,sc)]\}, \tag{5.17}$$

$$F^{out}(sc) = \cup_{t \in T} \{sr \mid sr \in RO(t), \forall_{sr^* \in RO(t)}[d^{out}(sr^*,sc) \leq d^{out}(sr,sc)]\}. \tag{5.18}$$

Based on these farthest scan register sets, we deduce the number of times a scan register sr in scan chain sc is used as a farthest input scan register during the application of all tests in **T**. A similar number can be given for the farthest output scan registers. Note that differing from (5.15) and (5.16) these numbers, respectively denoted as $N^{fin}(sc,sr)$ and $N^{fout}(sc,sr)$, depend on the configuration of the scan registers in the scan chain sc.

For $1 \leq i \leq n$

$$N^{fin}(sc,sr) = \Sigma_{t \in T} N(t) \cdot MaxApply(sr,t), \tag{5.19}$$

$$N^{fout}(sc,sr) = \Sigma_{t \in T} N(t) \cdot MaxObserve(sr,t), \text{ where} \tag{5.20}$$

$$\text{MaxApply}(sr,t) = 1 \text{ if } sr \in RA(t) \text{ and } \forall_{sr^* \in RA(t)}[d^{in}(sr^*,sc) \leq d^{in}(sr,sc)],$$
$$= 0 \text{ else;}$$

$$\text{MaxObserve}(sr,t) = 1 \text{ if } sr \in RO(t) \text{ and } \forall_{sr^* \in RO(t)}[d^{out}(sr^*,sc) \leq d^{out}(sr,sc)],$$
$$= 0 \text{ else.}$$

Example of Scan Register Properties

Consider the following scan chain sc with four scan registers.

$sc = (sr_1, sr_2, sr_3, sr_4)$.

The following three tests are applied via this scan chain.

test t_1: $RA(t_1) = \{sr_1, sr_3\}$, $RO(t_1) = \{sr_3\}$, $N(t_1) = 6$,
test t_2: $RA(t_2) = \{sr_4\}$, $RO(t_2) = \{sr_2, sr_4\}$, $N(t_2) = 4$,
test t_3: $RA(t_3) = \{sr_2, sr_3\}$, $RO(t_3) = \{sr_1\}$, $N(t_3) = 5$.

The length of each scan register is as follows.

$l(sr_1) = 5$, $l(sr_2) = 8$, $l(sr_3) = 3$, $l(sr_4) = 6$.

These characteristics are indicated in Figure 5.6.

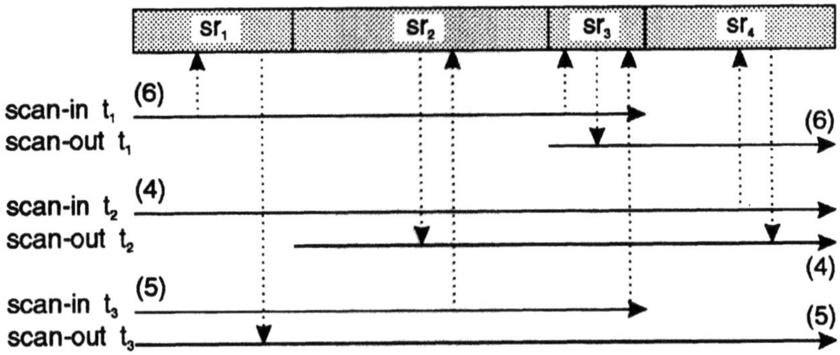

Figure 5.6. Example scan chain.

From (5.15) - (5.20) we can deduce the following figures for this example.

$N^{in}(sr_1) = 6$, $N^{out}(sr_1) = 5$, $N^{fin}(sc,sr_1) = 0$, $N^{fout}(sc,sr_1) = 5$,
$N^{in}(sr_2) = 5$, $N^{out}(sr_2) = 4$, $N^{fin}(sc,sr_2) = 0$, $N^{fout}(sc,sr_2) = 4$,
$N^{in}(sr_3) = 11$, $N^{out}(sr_3) = 6$, $N^{fin}(sc,sr_3) = 11$, $N^{fout}(sc,sr_3) = 6$,
$N^{in}(sr_4) = 4$, $N^{out}(sr_4) = 4$, $N^{fin}(sc,sr_4) = 5$, $N^{fout}(sc,sr_4) = 0$.

With (5.17)-(5.20) we can achieve our target to rewrite the cost function (5.9) for a scan chain configuration scc with a single scan chain sc. Instead as a sum over all tests, the cost function is now expressed as a sum over all scan registers in the $F^{in}(sc)$ and $F^{out}(sc)$ sets. This yields

$$C(scc) = \Sigma sr \in F^{in}(sc) \; [\; N^{fin}(sc,sr) \cdot d^{in}(sr,sc) \;] + \Sigma sr \in F^{out}(sc) \; [\; N^{fout}(sc,sr) \cdot d^{out}(sr,sc) \;]. \quad (5.21)$$

Since $N^{fin}(sc,sr) = 0$ if $sr \notin F^{in}(sc)$ and $N^{fout}(sr) = 0$ if $sr \notin F^{out}(sc)$ (5.21) can also be written as

$$C(scc) = \Sigma_{sr \in \mathbf{SR}} \; [\; N^{fin}(sc,sr) \cdot d^{in}(sr,sc) + N^{fout}(sc,sr) \cdot d^{out}(sr,sc) \;]. \quad (5.22)$$

Having rewritten the cost function (5.9) in terms of a sum over the scan registers instead as a sum of the tests, we now present the heuristics and local search approach for solving the Scan Chain Routing Problem SCRP for the special case of routing all scan registers in one scan chain.

Heuristics

To reduce the amount of shift operations, it is profitable for a scan register which applies many stimuli to be close to a scan chain input and for a scan register which observes many responses to be close to a scan chain output. This requirement is expressed in the *weight* of a scan register.

With the formulas (5.2), (5.15) and (5.16) the weight W(sr) for a scan register sr $\in$ **SR** is defined as follows.

$$W(sr) \; := \; [\; N^{in}(sr) - N^{out}(sr) \;] \; / \; l(sr). \quad (5.23)$$

Note that the weight of a scan register does not depend on the scan register position in a scan chain. The weight of a scan register is largely positive when the scan register is often used as applying register. On the other hand, it is largely negative when the scan register is often used as observing register.

We propose a heuristic approach based on a ranking according to decreasing weight [Oostdijk91]. As can be seen from Theorem 5.5, there are indeed cases where the minimum configuration is obtained when the registers are ranked according to decreasing weight.

Theorem 5.5

If $\forall_{t \in T} |RA(t)| = |RO(t)| = 1$, and (5.24)

$\cup_{t \in T} RA(t) = \mathbf{SR}$, $\cup_{t \in T} RO(t) = \mathbf{SR}$, (5.25)

then the configuration of scan registers in a single scan chain has minimal cost when the scan registers are sorted in order of decreasing weight from scan chain input to scan chain output.

Proof

Consider the scan chain configuration scc_1 with scan chain sc_1 consisting of n scan registers, $sc_1 = (sr_1, sr_2, ..., sr_n)$, with $sr_1 \rightarrow sr_2 \rightarrow ... \rightarrow sr_n$. Suppose that the scan registers in sc_1 are not configured in order of decreasing weight and that scc_1 is an optimal scan chain configuration. It is to be proven that there is a scan chain configuration scc_2 with $C(scc_2) < C(scc_1)$ and $scc_2 \neq scc_1$.

The scan registers in sc_1 are not configured in order of decreasing weight. This means that there are scan registers sr_i and sr_j in sc_1 such that $W(sr_j) > W(sr_i)$ and $i < j$. Without loss of generality we may state that $j = i+1$ since if there would be a scan register sr_q between sr_j and sr_i then either $W(sr_q) > W(sr_i)$ and then we can replace in our proof sr_j with sr_q, or $W(sr_q) \leq W(sr_i)$ but then we can replace sr_i with sr_q and meanwhile keeping the same property for the two considered scan registers.

From (5.24) and (5.25), we deduce that $F^{in}(sc_1) = F^{out}(sc_1) = \mathbf{SR}$ and each $sr \in \mathbf{SR}$ is used by exactly one $t \in \mathbf{T}$ as applying register and by exactly one $t \in \mathbf{T}$ as observing register. We therefore have the following relationship.

$$\forall_{sr \in \mathbf{SR}} \; [\; N^{in}(sr) = N^{fin}(sc_1,sr) \text{ and } N^{out}(sr) = N^{fout}(sc_1,sr) \;]. \tag{5.26}$$

From (5.2), (5.3) and (5.22), (5.26) we deduce that

$$C(scc_1) = \Sigma_{sr \in SR} [N^{in}(sr)\cdot[l(sr_1)+...+ l(sr)] + N^{out}(sr)\cdot[l(sr)+...+l(sr_n)]]. \tag{5.27}$$

We now construct the scan chain configuration scc_2 with scan chain sc_2 as follows from scc_1. If $sc_1 = (sr_1, ..., sr_i, sr_{i+1}, ..., sr_n)$ then $sc_2 = (sr_1, ..., sr_{i+1}, sr_i, ..., sr_n)$. From (5.24) and (5.25) we deduce that (5.26) holds also for scc_2. From this, it can easily be seen from (5.27) that

$$C(scc_1) - C(scc_2) = l(sr_i)\cdot l(sr_{i+1})\cdot[\ W(sr_{i+1}) - W(sr_i)\ ] > 0.$$

Hence, the cost of scc_2 is lower than the cost of scc_1, which is in contradiction with the hypothesis that scc_1 has minimal cost. □

Local Search

Another approach to solve the Scan Chain Routing Problem is to limit the search space by only looking to two consecutive elements in a scan chain and check whether it makes sense to interchange their positions. This approach is called *local search.* Our local search approach often requires an evaluation of a cost function difference. Using (5.22) we can express this cost function difference in terms of properties of two interchanged scan registers. This is the subject of the following Theorem.

Theorem 5.6

Let scc_1 be a scan chain configuration with the single scan chain $sc_1 = (sr_1, sr_2, ..., sr_i, sr_{i+1}, ..., sr_n)$ and scc_2 be the scan chain configuration with single scan chain sc_2 resulting of scc_1 by interchanging sr_i and sr_{i+1}, $sc_2 = (sr_1, sr_2, ..., sr_{i+1}, sr_i, ..., sr_n)$. Then

$$\begin{aligned} C(scc_1) - C(scc_2) &= d^{in}(sr_i,sc_1)\cdot[\ N^{fin}(sc_1,sr_i) - N^{fin}(sc_2,sr_i)\ ] \\ &+ d^{in}(sr_{i+1},sc_1)\cdot[\ N^{fin}(sc_1,sr_{i+1}) - N^{fin}(sc_2,sr_{i+1})\ ] \\ &+ d^{out}(sr_i,sc_1)\cdot[\ N^{fout}(sc_1,sr_i) - N^{fout}(sc_2,sr_i)\ ] \\ &+ d^{out}(sr_{i+1},sc_1)\cdot[\ N^{fout}(sc_1,sr_{i+1}) - N^{fout}(sc_2,sr_{i+1})\ ] \\ &- l(sr_{i+1})\cdot[\ N^{fin}(sc_2,sr_i) - N^{fout}(sc_2,sr_i)\ ] \\ &+ l(sr_i)\cdot[\ N^{fin}(sc_2,sr_{i+1}) - N^{fout}(sc_2,sr_{i+1})\ ]. \end{aligned} \tag{5.28}$$

Proof
Using (5.22) we derive the following expression for the difference in cost.

$$C(scc_1) - C(scc_2) = \Sigma_{sr \in SR} [\, N^{fin}(sc_1,sr) \cdot d^{in}(sr,sc_1) + N^{fout}(sc_1,sr) \cdot d^{out}(sr,sc_1) \,]$$

$$- \Sigma_{sri \in SR} [\, N^{fin}(sc_2,sr) \cdot d^{in}(sr,sc_2) + N^{fout}(sc_2,sr) \cdot d^{out}(sr,sc_2) \,]. \quad (5.29)$$

Since $N^{fin}(sc_1,sr_k) = N^{fin}(sc_2,sr_k)$ and $N^{fout}(sc_1,sr_k) = N^{fout}(sc_2,sr_k)$ for k = 1..n and k $\neq$ i,i+1, it follows from (5.29) that

$$C(scc_1) - C(scc_2) = [\, N^{fin}(sc_1,sr_i) \cdot d^{in}(sr_i,sc_1) + N^{fout}(sc_1,sr_i) \cdot d^{out}(sr_i,sc_1) \,]$$

$$+ [\, N^{fin}(sc_1,sr_{i+1}) \cdot d^{in}(sr_{i+1},sc_1) + N^{fout}(sc_1,sr_{i+1}) \cdot d^{out}(sr_{i+1},sc_1) \,]$$

$$- [\, N^{fin}(sc_2,sr_i) \cdot d^{in}(sr_i,sc_2) + N^{fout}(sc_2,sr_i) \cdot d^{out}(sr_i,sc_2) \,]$$

$$- [\, N^{fin}(sc_2,sr_{i+1}) \cdot d^{in}(sr_{i+1},sc_2) + N^{fout}(sc_2,sr_{i+1}) \cdot d^{out}(sr_{i+1},sc_2) \,]. \quad (5.30)$$

From (5.2), (5.3), and the special relationship between sc_1 and sc_2 we deduce that
$d^{in}(sr_i,sc_2) = d^{in}(sr_i,sc_1) + l(sr_{i+1})$, $\quad d^{out}(sr_i,sc_2) = d^{out}(sr_i,sc_1) - l(sr_{i+1})$,
$d^{in}(sr_{i+1},sc_2) = d^{in}(sr_{i+1},sc_1) - l(sr_i)$, $\quad d^{out}(sr_{i+1},sc_2) = d^{out}(sr_{i+1},sc_1) + l(sr_i)$.

Hence, we may write (5.30) as follows.

$$C(scc_1) - C(scc_2) = d^{in}(sr_i,sc_1) \cdot [\, N^{fin}(sc_1,sr_i) - N^{fin}(sc_2,sr_i) \,]$$

$$+ d^{in}(sr_{i+1},sc_1) \cdot [\, N^{fin}(sc_1,sr_{i+1}) - N^{fin}(sc_2,sr_{i+1}) \,]$$

$$+ d^{out}(sr_i,sc_1) \cdot [\, N^{fout}(sc_1,sr_i) - N^{fout}(sc_2,sr_i) \,]$$

$$+ d^{out}(sr_{i+1},sc_1) \cdot [\, N^{fout}(sc_1,sr_{i+1}) - N^{fout}(sc_2,sr_{i+1}) \,]$$

$$- l(sr_{i+1}) \cdot [\, N^{fin}(sc_2,sr_i) - N^{fout}(sc_2,sr_i) \,]$$

$$+ l(sr_i) \cdot [\, N^{fin}(sc_2,sr_{i+1}) - N^{fout}(sc_2,sr_{i+1}) \,].$$

□

A local search algorithm can be set up by each time comparing two consecutive elements in the scan chain and interchange their position if the cost difference function (5.28) is positive and keep them as they are if the cost difference function is negative or indifferent. An example of such a routine is as follows.

Routine SingleScanChainRoutingByLocalSearch;
sc_1 := (sr_1, sr_2, ..., sr_n);
scc1 := {sc_1} ;
i:=n-1;
while i $\neq$ 0
do begin
    sc_2 := (sr_1, sr_2, .., sr_{i-1}, sr_{i+1}, sr_i, sr_{i+2}, ..., sr_n);
    scc_2:={sc_2};
    compute sets F^{in}, F^{out}; compute numbers N^{fin}, N^{fout} for sr_i, sr_{i+1};
    if C(scc_1) - C(scc_2) > 0
    then **begin** scc_1:=scc_2;
        i:=min(i+1,n-1)
        end
    else i:=i-1;
end;
final_scan_chain_configuration := scc_1;

The initial scan chain routing is taken at random. Note that the pointer i does not jump back to position n-1 but to position min(i+1, n-1). This is due to the fact that the cost function difference can be expressed in terms of consecutive scan register elements.

Example of Usage of the Local Search Algorithm

Consider the following scan chain configuration scc with scan chain sc with 6 scan registers with an initial random ordering given by

$$scc = \{sc\} \text{ and } sc = (sr_1, sr_2, sr_3, sr_4, sr_5, sr_6).$$

Three tests are applied via this scan chain with the following characteristics.

Test t_1: $RA(t_1) = \{sr_1, sr_2\}$, $RO(t_1) = \{sr_3, sr_5, sr_6\}$, $N(t_1) = 10$,
Test t_2: $RA(t_2) = \{sr_4, sr_5\}$, $RO(t2) = \{sr_2, sr_4\}$, $N(t_2) = 7$,
Test t_3: $RA(t_3) = \{sr_2, sr_3, sr_5\}$, $RO(t_3) = \{sr_1\}$, $N(t_3) = 5$.

Furthermore the length of each scan register if given as follows.

$l(sr_1) = 5$, $l(sr_2) = 16$, $l(sr_3) = 5$, $l(sr_4) = 3$, $l(sr_5) = 3$, $l(sr_6) = 5$.

From this, (5.15)-(5.16), and (5.19)-(5.20) we can deduce the following values for this configuration of scan registers in this scan chain.

$N^{fin}(sc,sr_1) = 0$, $N^{fout}(sc,sr_1) = 5$, $N^{in}(sr_1) = 10$, $N^{out}(sr_1) = 5$,
$N^{fin}(sc,sr_2) = 10$, $N^{fout}(sc,sr_2) = 7$, $N^{in}(sr_2) = 15$, $N^{out}(sr_2) = 7$,
$N^{fin}(sc,sr_3) = 0$, $N^{fout}(sc,sr_3) = 10$, $N^{in}(sr_3) = 5$, $N^{out}(sr_3) = 10$,
$N^{fin}(sc,sr_4) = 0$, $N^{fout}(sc,sr_4) = 0$, $N^{in}(sr_4) = 7$, $N^{out}(sr_4) = 7$,
$N^{fin}(sc,sr_5) = 7$, $N^{fout}(sc,sr_5) = 0$, $N^{in}(sr_5) = 7$, $N^{out}(sr_5) = 10$,
$N^{fin}(sc,sr_6) = 5$, $N^{fout}(sc,sr_6) = 0$, $N^{in}(sr_6) = 5$, $N^{out}(sr_6) = 10$.

From (5.23) we can compute the weight of each scan register.

$W(sr_1) = 1$, $W(sr_2) = 0.5$, $W(sr_3) = -1$, $W(sr_4) = 0$, $W(sr_5) = -1$, $W(sr_6) = -1$.

Via the heuristic approach of arranging scan registers according to decreasing weight, the resulting scan chain would have been

$$sc = (sr_1, sr_2, sr_4, sr_3, sr_5, sr_6).$$

We now swap two consecutive scan register elements in the initial random scan chain configuration and use (5.28) to compute the difference in cost. Based on the new configuration, the new F^{in} set and F^{out} set can be easily computed including the numbers N^{fin} and N^{fout}. These new sets and numbers determine the final cost difference.

$sr_5 \leftrightarrow sr_6$	$\Delta C = -20$.	Cost increasing swap.	Reject.
$sr_4 \leftrightarrow sr_5$	$\Delta C = 0$.	Indifference swap.	Reject.
$sr_3 \leftrightarrow sr_4$	$\Delta C = 30$.	Cost decreasing swap.	Accept.

New scan chain, identical to the one derived via the heuristics
$(sr_1, sr_2, sr_4, sr_3, sr_5, sr_6)$, with
$F^{in} = \{sr_2, sr_5, sr_6\}$, $F^{out} = \{sr_1, sr_2, sr_3\}$.

$sr_3 \leftrightarrow sr_5$	$\Delta C = 35$.	Cost decreasing swap.	Accept.

New scan chain

$(sr_1, sr_2, sr_4, sr_5, sr_3, sr_6)$, with
$F^{in} = \{sr_2, sr_5, sr_6\}$, $F^{out} = \{sr_1, sr_2, sr_5\}$.

$\mathbf{sr_3 \leftrightarrow sr_6}$	$\Delta C = 0.$	Indifference swap.	Reject.
$\mathbf{sr_4 \leftrightarrow sr_5}$	$\Delta C = -30.$	Cost increasing swap.	Reject.
$\mathbf{sr_2 \leftrightarrow sr_4}$	$\Delta C = -30.$	Cost increasing swap.	Reject.
$\mathbf{sr_1 \leftrightarrow sr_2}$	$\Delta C = 45.$	Cost decreasing swap.	Accept.

New scan chain

$(sr_2, sr_1, sr_4, sr_5, sr_3, sr_6)$, with
$F^{in} = \{sr_1, sr_5, sr_6\}$, $F^{out} = \{sr_1, sr_2, sr_5\}$.

$\mathbf{sr_1 \leftrightarrow sr_4}$	$\Delta C = -15.$	Cost increasing swap.	Reject.

All consecutive scan register elements are now compared and no further decrease of cost function can be obtained via this local search algorithm. This gives the following final scan chain resulting from this local search routine.

$(sr_2, sr_1, sr_4, sr_5, sr_3, sr_6)$.

Table 5.2 shows the cost for the random initial scan chain configuration, the scan chain configuration derived from the weight-ordering heuristic approach, and the scan chain configuration derived from the local search algorithm.

Scan chain configuration	Cost
Initial scan chain configuration	1188
heuristic scan chain configuration	1158
local search scan chain configuration	1113

Table 5.2. Cost for the various scan chain configurations from the example.

6 Test Control Block Concepts

6.1 Introduction

The process of designing digital ICs is so complex that only by adopting a structured design methodology correct designs result within the constraints of time and financial budget. For this reason, synchronous design styles have emerged where often a distinction is made between a data path and a control path. The control path is needed to control the operation of the data path.

We have followed an identical reasoning for leaf-macro test data and test control access by considering a *test data path* and a *test control path*. We differentiate between functional paths and test paths by referring to the *functional data and control path* and the *test data and control path*. As explained in Chapter 3, the test data path may make use of the existing data path via the use of transfer properties. The transfer properties are controlled via transfer condition signals. In turn, the condition signals have to be controllable. The controllability of the transfer condition signals is handled as an item for the test data path. In cases where test data access cannot be provided via the functional data path, additional design entities are needed to enforce access. The control over these design entities is done via additional test control signals. These test control signals constitute the *test control path*. The combination of a leaf-macro test data path and a leaf-macro test control path is called the *leaf-macro access path*.

All test control signals have to be controllable during both test operation and normal operation. During test operation, the data to be applied to the test control signals has to follow the protocol stated in the leaf-macro generated test plans. During normal functional operation, the test control signals have to allow functional device operation. There are various ways to implement the test control path. For example, the functional control path can be extended, the test control signals can be connected directly to device pins, or a separate test control unit is used. To guarantee that the test control path does not interfere with the functional data and control path, and to adhere to a manageable and well defined design methodology, we have taken the approach to keep the test control path separate from the functional control path.

Normally, the test control signals are connected to additional device pins. However, the flexibility which Macro Test provides in test data path choices may lead to a complicated test control path. This may result in an unacceptable number of additional device pins. The solution to the pin count problem is to incorporate device specific test control logic in the device. This test controller is called a *Test Control Block* (TCB).

This chapter focuses the attention on the architecture and design of a *Test Control Block*. The Test Control Block is the 'conductor' of the device during test operation, whereas the functional control unit is the 'conductor' of the device during functional operation. Items to be discussed are the choice for a basic Test Control Block architecture, the Test Control Block design method, and optimization techniques to reduce the Test Control Block silicon area. Further, we shortly consider the issue of multi-layer test control. We limit our analysis to a single Test Control Block per device. At the end of this chapter, some statements are made on a multi-layer structure of Test Control Blocks.

6.2 Test Control Block Requirements

The *function* of a Test Control Block is to generate the required test control signal value sequences. The value sequences are defined in the leaf-macro generated test plans and may differ per design. This design dependency puts a requirement on the flexibility of a Test Control Block architecture and implementation. Further, the prime driving force for the use of a Test Control Block is limiting the number of test control device pins. Hence, a Test Control Block itself may not use too many device pins. Obviously, a Test Control Block should occupy a minimum of silicon area. Finally, the whole trajectory from Test Control Block specification to implementation should be automated and the corresponding software tools and cell library should be available in a design system as a module compiler.

6.3 Test Controller Architectures

Generation of controllers in general is a finite state machine synthesis problem. Fixed algorithms can be mapped onto a controller architecture via logic synthesis techniques. A lot of effort has been spent on suitable controller architectures [Zegers90]. The decision which test controller architecture to choose for our Macro Test purposes, is based upon the stated list of requirements and an analysis of published and used Test Control Block structures.

Examples of Test Control Block Structures

Conventional micro-controllers make use of a ROM for containing the control settings and some circuitry for counting or branching through the test control words. An example of a micro-programmed Test Control Block is described in [Breuer88]. The conclusions of Breuer are that if the design already has a microprogram control unit, then the test control words require little additional effort to be generated by the same control unit. However, if the design does not have a micro-programmable control unit, other test controller structures may be more desirable.

Most of the leaf-macro test data access methods are based on scan design techniques. The idea to provide test control also via a serial shift register and decode the instruction is therefore logical. These techniques automatically result in a multi-layer test control structure. A first layer is used to control the scanning process, and a second layer is used to apply the scanned test control words. We discuss four examples of such approaches.

A first example is given in [Riessen89]. A multi-layer structure of test control is described where control signals, which are only needed locally, are shifted to the appropriate places and properly decoded; see Figure 6.1.

The operating mode of the data scan cells is controlled by the test control signals; mode, drc1, and drc2. These test control signals are applied in parallel and are generated by a local test controller. The data scan cells can operate in four different modes; a hold mode in which data in the flip-flop remains unchanged, a test mode for Built-In Self-test purposes, a scan mode, and a normal functional mode. During the normal functional mode, the scan cells are transparent. The test control signals provided by the register cells M and S are controlled via a serial scanning process. An IC level test processor supplies the test control signals update_ir, irc1, and irc2 in parallel.The contents of the flip-flop L2 is loaded into the output latch L1 under control of update_ir. The signals irc1 and irc2 control which input is selected by the multiplexer. The select register S controls the bypass of the data scan cells. The control register M determines together with the parallel control signals from the IC-level test controller the function of the multiplexer of a data-register cell.

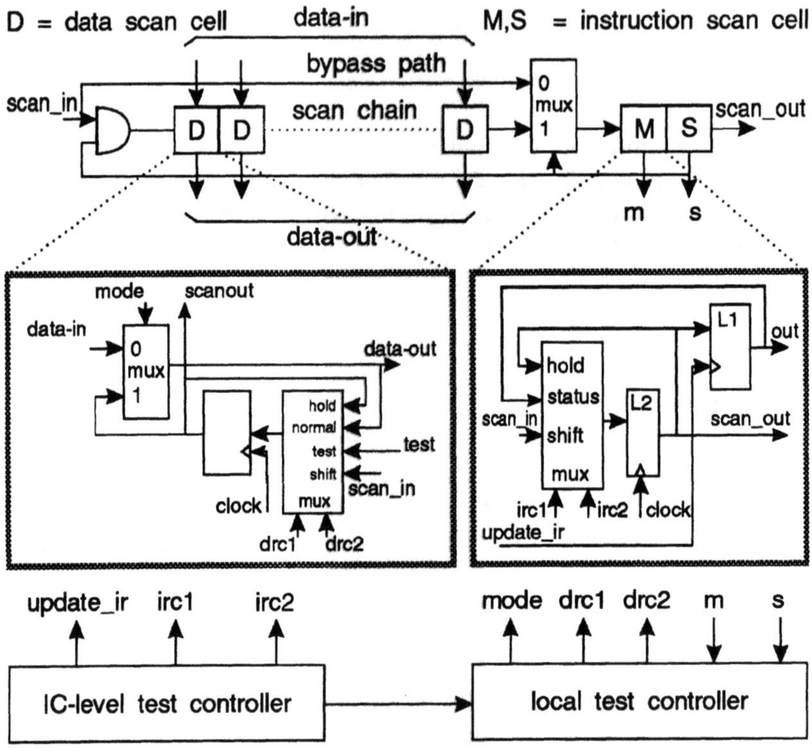

Figure 6.1. A multi-layer test control structure.

The approach as described in [Riessen89] is based on standard leaf-macro access mechanisms and supports leaf-macro Built-In Self-Test. The standard access mechanism requires each leaf-macro to be surrounded by the data scan cells, control register cells, and a local test controller. The disadvantage is that it is not flexible enough to support a variety of test control value sequences for controlling other test data paths than scan for the leaf-macros. The layered structure of test control is carefully tuned and works due to the standard scan access. Further, the requirements for the scan cells and controller around each leaf-macro has a large impact on the use of additional silicon real estate.

A second example of a two-layer test control structure applying serial test control access is described in [Leenstra93a, Leenstra93b]. The serial test control is used for bypassing portions of scan chains to obtain a fast access to certain leaf-macros; see Figure 6.2.

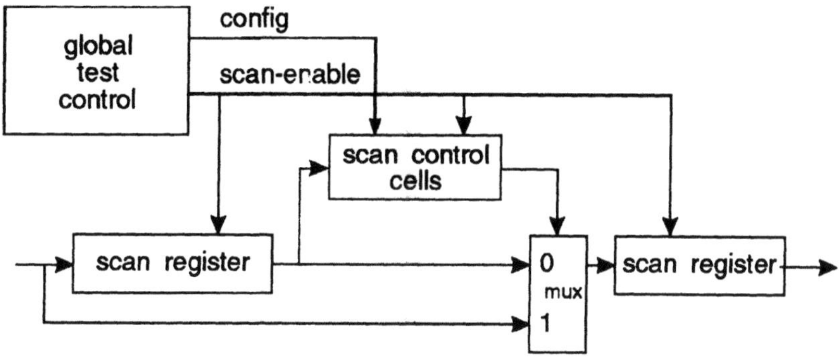

Inputs		Operation modes	
scan_enable	config	scan register	scan control cells
0	0	normal	hold
0	1	normal	update
1	0	serial scan	hold
1	1	serial scan	reset

Figure 6.2. Two-layer test control structure for reconfiguring scan chains.

The shift process of the test control precedes the shift process of the test data and is itself controlled by a two device pins, called scan-enable and config. The configuration of the scan chain is done as follows. First, the scan-enable and config signals are set to '1'. This causes a reset of the scan control cells. All multiplexers are set to the '0' state and the configuration pattern is scanned into the scan registers. The scan control cells hold their '0' value during this process. To store the scan register data into the scan control cells, the scan-enable signal is first set to '0'. Thereafter, the config signal is set to '1', which causes the propagation of the scan control cell input data to their output. As a result, the multiplexer control inputs are set to the value as defined by the configuration pattern. The scan control cell must hold its value during the execution of the test patterns, so next the config signal is set back to '0'.

The approach as described by Leenstra requires two additional device pins and is based on scan access to leaf-macros combined with a standard control structure.

Another example of a shift register test control approach is described in [Hapke89]. A serial shift register defines a control word which is decoded and fixed for a certain period of time in a test. The leaf-macro test data path is standardized on serial scan and the decoded instructions determine the values of the test control signals during the scan mode, the apply cycle, and normal functional operation. The number of instructions which can be decoded is limited and at maximum two control words are available during a leaf-macro test. The decoding possibilities are stored in a the design library and handled by a TCB module compiler.

The final example we discuss is the Boundary Scan standard [IEEE90]. This is an example of a Test Control Block where only one device pin (TMS) is used for test control purposes and where a multi-layer control structure is applied. The Boundary Scan architecture consists of a series of scannable data registers, a scannable instruction register, and a Test Access Port (TAP) test controller; see Figure 6.3. The data in the instruction register is responsible for the selection and test mode of one of the other data registers. The finite state machine in the Test Access Port controller is responsible for the control required to shift data through either the instruction register or one of the data registers. This controller itself is controlled by the signal TMS.

The Boundary Scan architecture has a multi-layer test control structure. The TAP controller is the dominant controller and provides *parallel* test control to the instruction register and the other data registers. The instruction register is the second level test controller which receives its data *serially* and provides parallel test control data to set up the test data path through the data registers.

In all four examples, there is a clear distinction between the control over the scanning process and the set-up of the test data path. All four examples rely on a full scan approach and make use of a *standardized* controller structure. In Section 6.8, we argue that a multi-layer test control is only feasible in case of a standardized test data and control path. This is in contradiction with the flexibility of the Macro Test approach which allows a design-specific choice for test data paths. The choice for test data path determines the test control path.

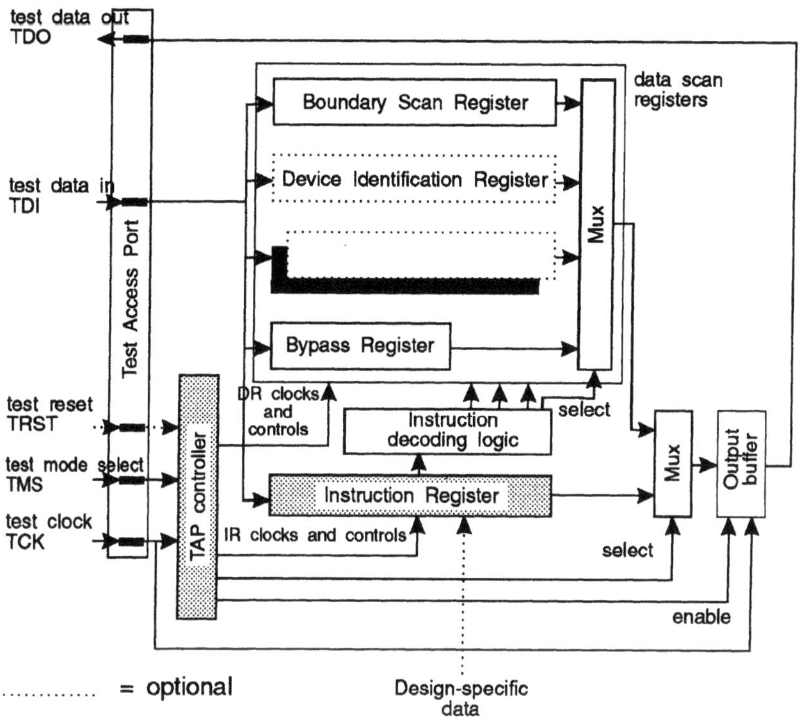

Figure 6.3. The Boundary Scan architecture.

We have made a choice for a test control approach which allows flexibility in choice for test data paths. The leaf-macro test data paths, including the required test control value sequences, are stated in the leaf-macro generated test plans. The function of a Test Control Block is to generate the test control value sequences. A straightforward Test Control Block approach is to map these sequences onto a single finite state machine controller, making use of well-known logic synthesis techniques [Bouwman92b]. Every clock cycle, this finite state machine produces a full control word. To limit the number of device pins, the choice is to control this finite state machine with one device pin; see Figure 6.4.

This Test Control Block approach has been used to our satisfaction in the design of several ICs. All application examples had a relatively small number of test control words and test control signals. However, for more complicated test plans with a large number of control words, a ROM based approach might be more appropriate. We limit ourselves to one finite state machine Test Control Block per design.

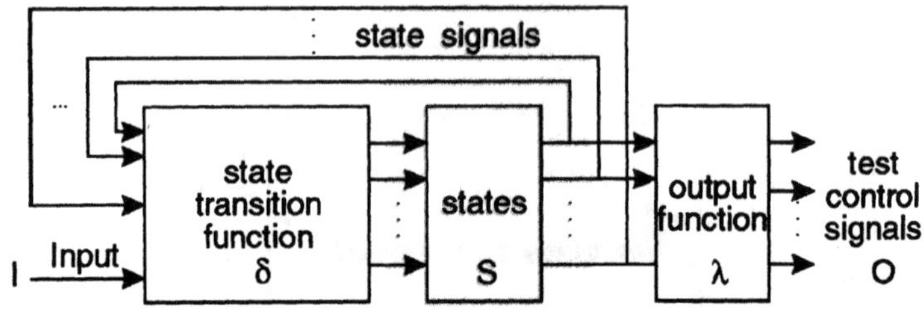

Figure 6.4. Single input finite state machine TCB. δ is the state transition function, λ is the output function.

6.4 Relation between a Test Control Block and Test Plans

Test Plan Mapping

Every leaf-macro has an associated generated test plan. Among other things, a generated test plan contains all information about the value sequences to be applied to the test control ports. These value sequences are given in chronological order as *test control words*.

A finite state machine is used to implement the test control structure. States and the state transition function δ realize the *order* in which the test control words have to be applied. The *content* of the test control words is derived from the states by the output function λ; see Figure 6.4. A *mapping* of the *sequence* and *content* of test control words onto a state diagram, considering the state transition function δ and output function λ separately, is straightforward. Such a state diagram is called a *leaf-macro state diagram* (LMSD). An example is given in Figure 6.5. The test control section of the generated test plan is highlighted. Each test control word is represented by a state. The states are indicated by a circle and arrows denote the transition from one state into another. Decisions have to be taken under which condition a transition occurs. The order of the test control words (δ) and the content of the test control words (λ) are indicated in the leaf-macro state diagram.

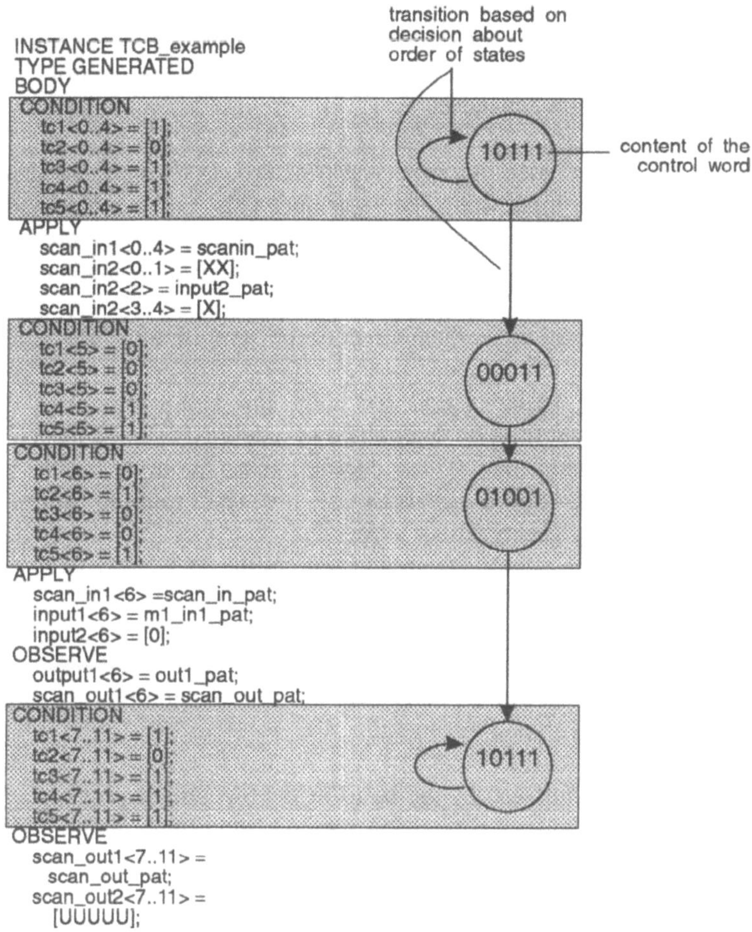

Figure 6.5. Example of mapping a sequence of test control words onto a leaf-macro state diagram.

The sequence of leaf-macro test control words control the operation of the leaf-macro test data path. Test control signals which are not used for a certain leaf-macro, have a don't care value.

Test Control Block Assembly

The process of test plan mapping is repeated for every leaf-macro in the design, resulting in a set of individual leaf-macro state diagrams. The leaf-macro test data path may make use of data paths through other leaf-macros or leaf-macros may share test data paths. Hence, leaf-macros may have similar test control words and similar test control word sequences.

Having the leaf-macro state diagrams, the next task is to *assemble* these state diagrams into a single input finite state machine. This state machine constitutes the Test Control Block functionality and is called the *TCB finite state machine* (TCB FSM). The assembly process needs to satisfy the following conditions.

* Each leaf-macro state diagram must be part of the TCB finite state machine.
* The decision under which condition the leaf-macro state diagram transitions occur must be determined by a defined value sequence applied to the TCB input port. These decisions must be in according with the sequence and timing intervals of the test control words in the corresponding generated test plan.
* Transitions are possible from any state in the TCB FSM to any state in the leaf-macro state diagram.
* During normal operation of the device, the Test Control Block must be fixed into a pre-defined state. This state is called the *TCB reset state*. The reset state allows normal functional device operation by forcing all TCB output signals to non-test mode.

These conditions are fulfilled by the basic TCB FSM structure, given in Figure 6.6. This structure provides a well visible relation between the TCB FSM and the leaf-macro state diagrams.

Three parts can be distinguished in Figure 6.6. A reset part, a selection part and the test plan part. If the IC is operating in normal mode, the TCB FSM must be looping in the reset state which generates the test control words required for normal circuit operation. The selection part enables the selection of a generated test plan for each leaf-macro. The test plan part consists of the leaf-macro state diagrams resulting from the test plan mapping process.

The structure of the TCB FSM shown in Figure 6.6 forms a basic concept and is very similar to the structure of the finite state machine controller in the Boundary Scan architecture [IEEE90]. However, a structure itself is not enough. Design requirements such as minimal use of silicon real estate have been stated. The complexity of the TCB FSM has to be handled and the state machine has to be efficiently mapped onto a final gate level specification. In the following sections, we make a choice for a structured design model and we consider various silicon-area optimization techniques.

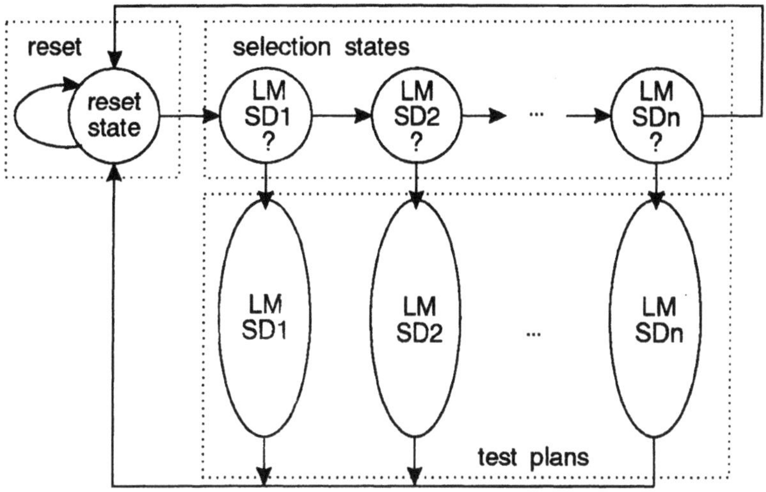

Figure 6.6. The basic Test Control Block finite state machine structure.

6.5 Test Control Block Design Requirements

The most important Test Control Block design requirements can be formulated as follows.

* All output transitions have to be initiated by a clock and have to take place within a certain timing interval.
* Only one input port is available for controlling the TCB.
* After a power-on procedure of the device there must be a guaranteed and easy way to enter the reset state of the Test Control Block.
* In the reset state of the Test Control Block, the circuit performs its normal functional operation.

The TCB is a sequential machine. This means that the behaviour of the TCB is influenced by former events. In the literature two classes of finite state machines are distinguished: *Mealy machines* and *Moore machines* [Mealy55, Moore64, Thijssen90]. In Mealy machines, there is a direct coupling between input and output signals, see Figure 6.7. As a result, the timing of the output signals of a Mealy machine is not only determined by the functional model of the design but also by the timing of the input signals. Combinatorial loops may exist as a result of which oscillations may occur.

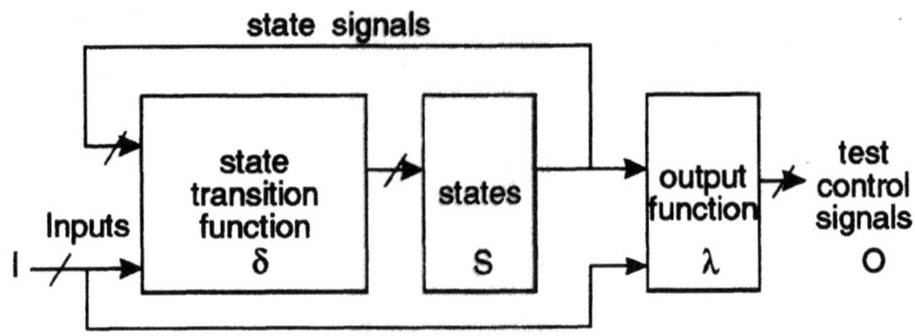

Figure 6.7. Block diagram of a Mealy Machine.

In Moore machines, there is no direct coupling between input and output. Output transitions are always initiated by a clock pulse and bound to a certain timing interval; see Figure 6.8. Mathematically speaking, the Moore machine is a special case of the Mealy machine [Hartmanis66]. To circumvent the risks of a Mealy machine we have made the choice to implement the TCB as a Moore machine.

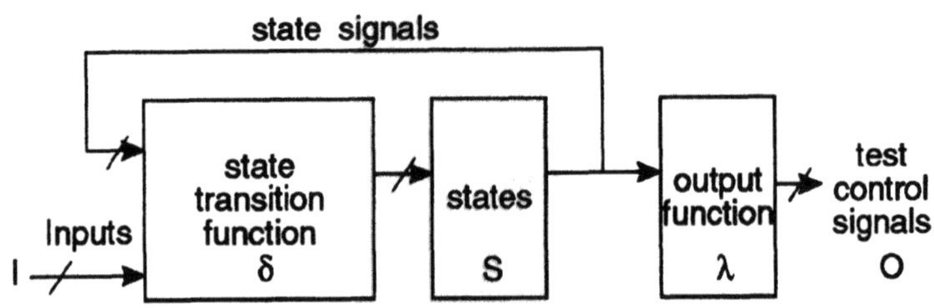

Figure 6.8. Block diagram of a Moore Machine.

A Moore machine M can be formalized as a five tuple of the form

$$M = (I,O,S,\delta,\lambda),$$

where

- I : a finite, non-empty set of inputs,
- O : a finite, non-empty set of outputs,
- S : a finite, non-empty set of internal states,
- δ IxS $\rightarrow$ S : the state transition function,
- λ S $\rightarrow$ O : the output function.

A finite state machine specification does not usually have a number of states equal to a power of two. Hence, if the TCB is implemented by means of binary flip-flops, a number of states is unused. For these unused states both the state transition function and the output function have to be specified also. If not, then a power-on procedure may initialize the TCB in one of the unspecified state resulting in undefined behaviour. A finite state machine implementation without unspecified behaviour for unused states is called a *fully specified finite state machine*.

The first design requirement, only clocked transitions bound within a certain timing interval, is basically met by implementing the TCB as a Moore Machine. This guarantees that no output transition occurs which is not initialized by a clock pulse. It therefore does not have the risk that a transition on the input signal results in undesired actions of the Test Control Block. The requirement that an output transition takes place within a certain timing interval is still a point of concern and puts a requirement on the design. This can be solved by carefully selecting the design library elements. Special attention must then be given to timing aspects of flip-flops and other sequential elements; see Chapter 8.

The requirement of only one input port allows only two next states per current state. The TCB device input pin is called the *MTM* pin (Macro Test Mode). By a careful input value assignment to the MTM pin we can guarantee that the reset state is always reached by keeping MTM active high. Reaching of the reset state by keeping MTM active high is called the *reset requirement*. The *input assignment* problem has been solved with algorithms derived from graph theory and is described in [Marinissen90, Marinissen91]. The Test Control Block input assignment problem is similar to the reachability problem in directed graphs. An example of a TCB FSM input assignment, which fulfils the reset requirement, is given in Figure 6.9. The input assignment is a straightforward process. For this reason, we have not indicated the input assignment in the example TCB finite state machines in the rest of this chapter.

The reset requirement puts restrictions on unused states. There are ways of specifying unused states, which lead to an optimal gate level implementation of the Test Control Block; see section 6.6.3. We only deal with fully specified finite state machines, therefore meeting the requirement on reaching the reset state from any other used or unused state.

The final design requirement, a normal functional behaviour in the reset state, puts a requirement on the output function λ. The output function for the reset state is such that the test control signals allow normal functional operation.

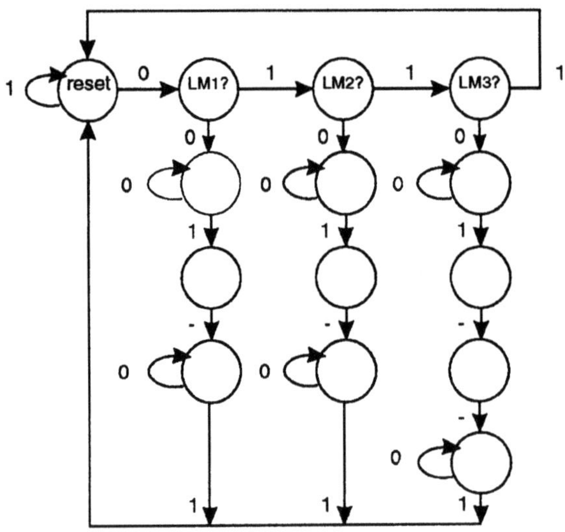

Figure 6.9. Example of a TCB FSM input assignment.

6.6 Optimal Test Control Block implementation

The presented structure of a Test Control Block forms a basic concept. The next step is to map this structure onto a gate level implementation via state assignment and logic synthesis. As in IC design practice, the objective is to obtain an optimal implementation. Optimal with respect to a minimal number of logic gates, which translates in minimal use of Test Control Block silicon area. Optimal also with respect to correct static and dynamic behaviour of the TCB circuit. A correct static behaviour is guaranteed by the model and by realizing the model in the circuitry. A correct dynamic behaviour needs more attention. We have stated as design requirement that an output transition takes place within a certain timing interval. This does not guarantee that the dynamic behaviour of all output signals is identical. To circumvent problems with dynamic properties of the output signals we have decided to place a register at the output of the Test Control Block. This guarantees that all output signals are available at the same time; see Figure 6.10.

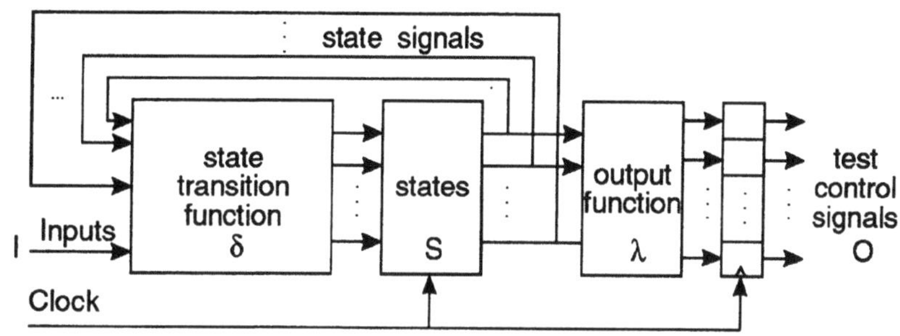

Figure 6.10. A register placed at the TCB output.

Looking at the TCB structure, we identify the following two possible ways to optimize the TCB implementation.

* Optimization with respect to a minimal number of finite state machine states.
* Optimization with respect to a minimal number of logic gates necessary to implement the state transition function and the output function.

In other words, the first optimization method focuses the attention on the flip-flop part of the TCB circuit. The second optimization method focuses the attention on the logic part of the TCB circuit. Both techniques have been investigated to some extend. These limited studies, described in [Marinissen91] and [Bouwman92b], show that the second technique is to be preferred above the first technique because of the basic structure of the TCB FSM. The main reason is that the state minimization leads to a complex state transition function and, therefore, to a relatively expensive gate level implementation, while saving only a few flip-flops.

A study of minimization of logic gates required to implement the Test Control Block state transition function and output function, taking into account the TCB basic structure, involves an elaboration of general logic minimization techniques, such as state merging, state assignment, and specification of unused states. The techniques of state merging, state assignment, and specification of unused states are based on the notion of equivalent and compatible states. Hence, before we can elaborate on these techniques we need some background on state equivalence and state compatibility. For this, we follow the definitions given in [Thijssen90, Thijssen93].

Let I be the set of input symbols of a sequential Moore machine, and let I^* be the set of all finite input sequences that can be constructed by concatenating the elements of I in any allowable order. An input sequence $J \in I^*$ is an ordered string of input symbols, denoted by $J = [i_1\ i_2\ ...\ i_n]$.

The successor state $\delta^*(s, J)$ of state $s \in$ S under input sequence J is the state in which the sequential Moore machine is when this machine starts in s and successively all input symbols of J are applied, beginning with i_1, i_2, ... The extended state transition function δ^* maps S x I^* into S.

In the same way the extended output function $\lambda^* : S \times I^* \rightarrow O$ can be defined.

Definition 6.1. Two states s_i and s_j of a completely specified sequential Moore machine M are said to be *equivalent*, denoted by $s_i \equiv s_j$, if and only if $\lambda^*(s_i, J) = \lambda^*(s_j, J)$, for all $J \in I^*$

This equivalence definition induces an equivalence relation R on the set of internal states S of a sequential Moore machine. This equivalence relation R is reflexive ($\forall_{s \in S}$ sRs), symmetric ($\forall_{s,t \in S}$ sRt $\Rightarrow$ tRs) and transitive ($\forall_{s,t,u \in S}$ sRt and tRu $\Rightarrow$ sRu).

An equivalence relation on the set of internal states induces a partition of S into a set of disjoint equivalence classes. All pairs of states from an equivalence class are equivalent. This implies that equivalent states cannot be distinguished by any logical input-output experiment. Further, it is well known from sequential machine theory that this implies also that the successor states of any pair of equivalent states can never be distinguished by any logical input-output experiment [Paull59]. Hence, for two equivalent states s_i and $s_j \in$ S we have that

$s_i \equiv s_j \Rightarrow \delta(s_i, i) \equiv \delta(s_j, i)$ for all input symbols $i \in$ I, and as a consequence,

$$s_i \equiv s_j \Rightarrow \delta^*(s_i, J) \equiv \delta^*(s_j, J) \text{ for all J} \in I^*.$$

These conclusions on equivalent states may also be used to expand the state set to a sequential machine up to 2^n internal states by splitting states into as many equivalent states as necessary to obtain a number of 2^n states. The successor states of the newly introduced states must be defined in such a way that make them equivalent to states of the same equivalence class, that is $\delta(s, i) \equiv \delta(s_j, i)$ for all $i \in I$ and s in the equivalence class of s_j. This can be done by making $\lambda(s, i) = \lambda(s_j, i)$ for the present output of s and choosing $\delta(s, i)$ as an element of the equivalence class of $\delta(s_j, i)$.

Next, we define a compatibility relation of two states.

Definition 6.2.
Let $M_a = (I_a, O_a, S_a, \delta_a, \lambda_a)$ be an incompletely specified sequential Moore Machine. Two states $s_i, s_j \in S_a$ are *compatible* if and only if there is a sequential Moore Machine $M_b = (I_b, O_b, S_b, \delta_b, \lambda_b)$ such that M_b can replace M_a under all circumstances and there is one state in S_b which covers both s_i and s_j. Two states $s_i, s_j \in S_a$ are covered by one state $r \in S_b$ if

$$\lambda^*_b(r, J) = \lambda^*_a(s_i, J) \text{ for all } J \in I^*(s_i),$$
$$\lambda^*_b(r, J) = \lambda^*_a(s_j, J) \text{ for all } J \in I^*(s_j),$$

where $I(s)$ is the set of specified input symbols for state $s \in S_a$ and $I^*(s)$ is the set of all finite input sequences that can be constructed by concatenating the elements of $I(s)$ in any allowable order. Compatibility of two states is a reflexive, symmetric, but not necessarily a transitive relation. A compatibility study involves the creation of a new Moore machine.

6.6.1 TCB Optimization via State Merging

State properties, such as compatibility and equivalence, are traditionally used for reducing the number of internal states of a sequential machine. A detailed elaboration of state compatibility can be found in [Thijssen90, Marinissen91]. In the case of our Test Control Block design we call this process of reducing states *merging of states*. Two compatible or equivalent states can be merged into one.

The type of merging as introduced in [Marinissen91] focuses on the computation of the maximal mergeable subgraph, i.e., the maximum set of equivalent or compatible states of two leaf-macro state diagrams. It has been found by Bouwman and Thijssen [Bouwman92b, Thijssen93] that it is more profitable to focus the attention only on state merging within a leaf-macro state diagram and to perform merging of two states from different leaf-macro state diagrams only when the two leaf-macro state diagrams are completely equivalent, i.e., belong entirely to the same maximal mergeable subgraph. In this manner the original structure of the TCB is preserved and kept simple and regular.

6.6.2 TCB Optimization via State Assignment

The next step in the Test Control Block design is state assignment, which is the assignment of a binary code to every internal state. State assignment procedures have been studied extensively in literature. Different assignments may have a considerable effect on the amount of hardware required to implement the Test Control Block state diagram. We have studied the question whether we could make use of the basic structure of the TCB to derive an efficient state assignment procedure. The results of this study are described in [Bouwman92b, Thijssen93]. Here we summarize the major conclusions. We have taken the approach to consider the state transition function δ and the output function λ separately. The state assignment is used for implementing the state transition function. For the output function, a decoder is taken which produces the TCB outputs on basis of the given state assignment.

The TCB FSM structure of Figure 6.6 consists of three parts, i.e., the test plan states, the selection states, and the reset state. In our state assignment procedure, the reset state and the selection states are considered one part. We first consider the state assignment for each part separately. Based upon these observations we define the state assignment of the total TCB FSM.

Let us first consider the reset state and the selection states. As can be seen from Figure 6.6, these states form a cycle. State assignment of this part only, requires the following number of bits

$$\text{\# selection bits} = \lceil \log_2(\text{\# selection states} + 1) \rceil.$$

It is well known from state assignment theory that for a cycle of states a counting assignment works properly. Hence, we start with coding the reset state with all zero and use a binary count for each successor selection state. An example is given in Figure 6.11. In this example, there are 3 selection states. Including the reset state, there are 4 states to be assigned. This requires 2 bits for the state code. The reset state is coded '00', the following three selection states are successively coded via a binary count.

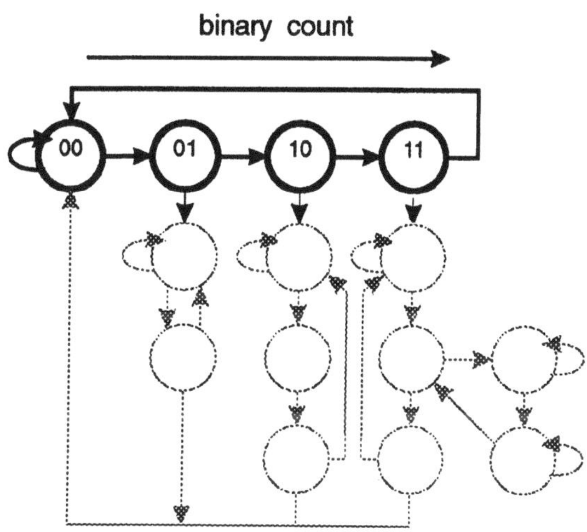

Figure 6.11. Example of a binary count state assignment of the reset state and selection states.

Next, we consider a state assignment for each leaf-macro state diagram, *including* its selection state. The number of bits required for such a state assignment is

$$\text{\#test plan bits} = \lceil \log_2(\text{\# leaf-macro state diagram states} + 1) \rceil.$$

Note that this number may be different for each leaf-macro state diagram.

Following an identical reasoning as for the state assignment for the reset state and selection states, we start with coding the selection state with all zero and use a binary count for each successor state in the leaf-macro state diagram; see Figure 6.12. In this example, there are 3 states in the leaf-macro state diagram. Including the selection state, there are 4 states to be assigned. This requires 2 bits for the state code. The selection state is coded '00', the following three leaf-macro states are successively coded via a binary count.

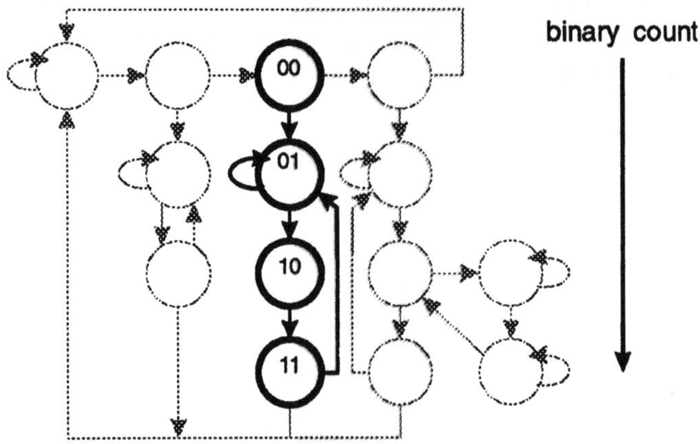

Figure 6.12. Example of a binary count state assignment of a selection states and test plan states.

The state encoding for the total Test Control Block state diagram is a combination of the two separate state assignments. We simply paste the two state assignments together in the following manner.

The number of bits used in the TCB FSM state assignment equals

$$\# \text{ selection bits} + \max \{tp_i \mid tp_i = \# \text{ test plan bits for leaf-macro } i, \; 1 \leq i \leq \# \text{ leaf-macros}\}.$$

The first section of each binary code is called the *selection code*. The remaining section is called the *test plan code*. The definition of the TCB FSM state assignment is as follows.

Selection code:

* The bits of the selection code for the selection states and the reset state are as given by its separate state assignment.
* The bits of the selection code of the test plan states are equal to the selection code of the corresponding selection state.

Test plan code

* The bits of the test plan code of the reset state are coded don't care. This means that a number of states are equivalent and together form the reset state.
* The bits of the test plan code for the selection states and the test plan states are coded equal to its separate state assignment. In the case more bits are required as necessary for its separate state assignment the unnecessary bits are coded don't care.

An TCB FSM state assignment example is given in Figure 6.13.

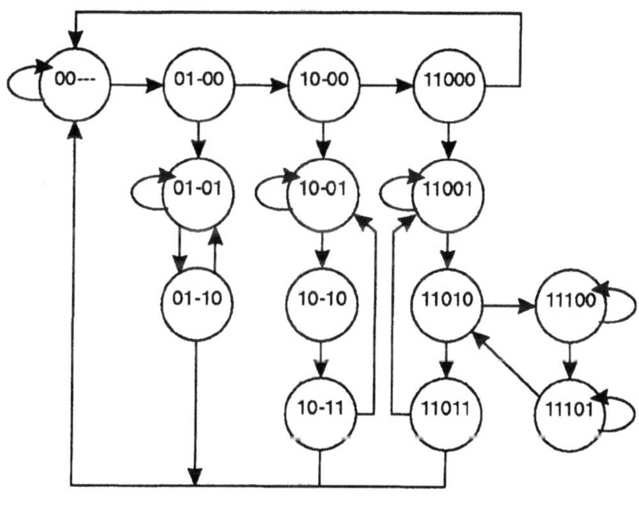

Figure 6.13. Example of a TCB FSM state assignment.

A further optimization can be done by improving the test plan state assignment. However, experiments described in [Bouwman92b] show that this state assignment produces good results.

Don't care values have to be treated carefully. They result in a number of unused states. Making use of state equivalence properties, we can map these states onto any other equivalent state. Which state to choose is a problem in itself and not treated in this book. To adhere to our choice of a systematic state assignment technique we propose to map an unused selection state onto the reset state and an unused test plan state onto a test plan state. A more detailed analysis of the use of unused states is given in the following section.

Another point of attention for unused states is the reset requirement. The reset requirement, stating that the Reset State is always reached by keeping the signal value of MTM active high, is met by a careful value assignment to the MTM port.

6.6.3 TCB Optimization via Specification of Unused States

The number of Test Control Block states depends on the number and sequences of control words that have to be generated for each leaf-macro generated test plan. This number is usually not a power of 2. This results in a number of unused states. The question is what to do with unused state codes which result after the state assignment procedure. This problem is treated by [Thijssen93] and is summarized in this section.

There are several options to handle unused states. We discuss the following ones.

Option 1

Keep the state transition function and output function for the unused states undefined.

$$\delta(s_d, i) = \text{'-'} \text{ for all input symbols } i \in I \text{ and unused states } s_d \in S,$$
$$\lambda(s_d) = \text{'-'} \text{ for all unused states } s_d \in S.$$

This has the advantage that logic synthesis tools can make use of the unspecified behaviour which may result in simple state transition and output equations. The drawback is that due to external disturbances the circuit may jump into an unspecified state or cycle of such states, which may create a dead-lock situation.

Option 2

Specify the state transition function and output function of an unused state s_d as a transition to the reset state s_R.

$$\delta(s_d, i) = s_R \text{ for all input symbols } i \in I, \text{ and unused states } s_d \in S,$$

$$\lambda(s_d) = \lambda(s_R) \text{ for all unused states } s_d \in S.$$

After a transition to an unused state the circuit is back into its normal cycle at the next clock pulse. Hence, the behaviour of the circuit under noisy conditions can never be a dead-lock situation.

Option 3

Expand the state set up to 2^n internal states. If all unused states s_d are specified as

$$\delta(s_d,i) = \delta(s_R,i) \text{ for all input words } i \in I, \text{ unused states } s_d \in S \text{ and reset state } s_R.$$

$$\lambda(s_d) = \lambda(s_R),$$

then unused states s_d behave as the reset state s_R. The external behaviour is identical which is a special case of state equivalence. As such the difference with option 2 is that a state equivalent to the reset state is effected immediately while with option 2 the reset state is only present after the next clock pulse.

Option 4

Extend option 3 as follows. A transition to the reset state may be changed into a transition to any other state that is equivalent to the reset state s_R. The same holds for transitions from any other state, which puts the circuit into normal functional operation, into the reset state s_R. As the successor for this state a jump into any successor state of the reset state s_R can be chosen.

Comparison experiments as described in [Thijssen93] have shown that option 4 is the preferred option for an optimal logic design of a Test Control Block.

6.7 Test Control Block Design Example

We consider a circuit consisting of four leaf-macros. The relevant conditional part of the four generated test plans is indicated in Figure 6.14. There are four test control signals, called tc1, tc2, tc3 and tc4. The signal value is indicated between square brackets ('['). The timing interval is indicated between angular brackets ('<').

```
Instance LeafMacro1
CONDITION tc1<0..4> = [1];
          tc2<0..4> = [0];
          tc3<0..4> = [1];
          tc4<0..4> = [1];
CONDITION tc1<5> = [0];
          tc2<5> = [1];
          tc3<5> = [0];
          tc4<5> = [0];
CONDITION tc1<6..11> = [1];
          tc2<6..11> = [0];
          tc3<6..11> = [1];
          tc4<6..11> = [1];
```

```
Instance LeafMacro2
CONDITION tc1<0..9> = [1];
          tc2<0..9> = [0];
          tc3<0..9> = [1];
          tc4<0..9> = [1];
CONDITION tc1<10> = [0];
          tc2<10> = [1];
          tc3<10> = [0];
          tc4<10> = [0];
CONDITION tc1<11..13> = [1];
          tc2<11..13> = [0];
          tc3<11..13> = [1];
          tc4<11..13> = [1];
```

```
Instance LeafMacro3
CONDITION tc1<0..7> = [1];
          tc2<0..7> = [0];
          tc3<0..7> = [1];
          tc4<0..7> = [1];
CONDITION tc1<8> = [0];
          tc2<8> = [0];
          tc3<8> = [0];
          tc4<8> = [0];
CONDITION tc1<9> = [0];
          tc2<9> = [1];
          tc3<9> = [0];
          tc4<9> = [0];
CONDITION tc1<10..15> = [1];
          tc2<10..15> = [0];
          tc3<10..15> = [1];
          tc4<10..15> = [1];
```

```
Instance LeafMacro4
CONDITION tc1<0..2> = [0];
          tc2<0..2> = [0];
          tc3<0..2> = [1];
          tc4<0..2> = [1];
CONDITION tc1<3> = [0];
          tc2<3> = [1];
          tc3<3> = [0];
          tc4<3> = [0];
CONDITION tc1<4..10> = [0];
          tc2<4..10> = [0];
          tc3<4..10> = [1];
          tc4<4..10> = [1];
```

Figure 6.14. Example of test control value sequences for four leaf-macros.

We follow the Test Control Block design process step by step.

Test Plan Mapping

Each leaf-macro generated test plan is mapped onto a state diagram. The result of this test plan mapping process is straightforward and is indicated in Figure 6.15. The output function of each state is given in the state circle. The output function of both the reset state and the selection states equals the reset value of the test control signals.

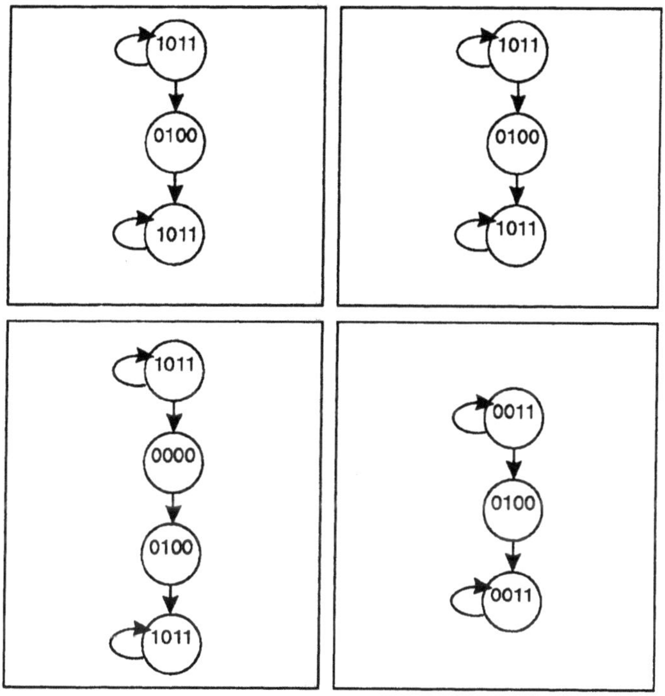

Figure 6.15. Test plan mapping resulting in four leaf-macro state diagrams.

Test Control Block Assembly

The leaf-macro state diagrams are assembled into a TCB FSM structure by adding the selection states and the reset state; see Figure 6.16.

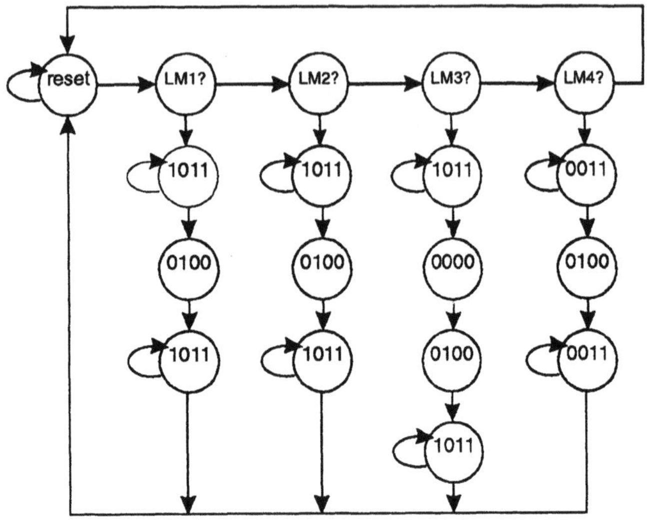

Figure 6.16. Example Test Control Block assembly result.

Test Control Block Implementation

The result of Test Control Block assembly is now ready to be implemented into a fully specified Moore machine via the steps of input assignment, state merging, and state assignment.

Input Assignment

The MTM signal values for our example Test Control Block are given next to each state transition arc in Figure 6.17. Notice, that indeed keeping the MTM signal value at a logical 1 fulfils the reset requirement.

State Merging

Leaf-macro state diagram 1 and leaf-macro state diagram 2 are equivalent. Hence, they can be merged into one leaf-macro state diagram. Secondly, there are some internal merging possibilities due to some compatible states. The result of this state merging process is indicated in Figure 6.18.

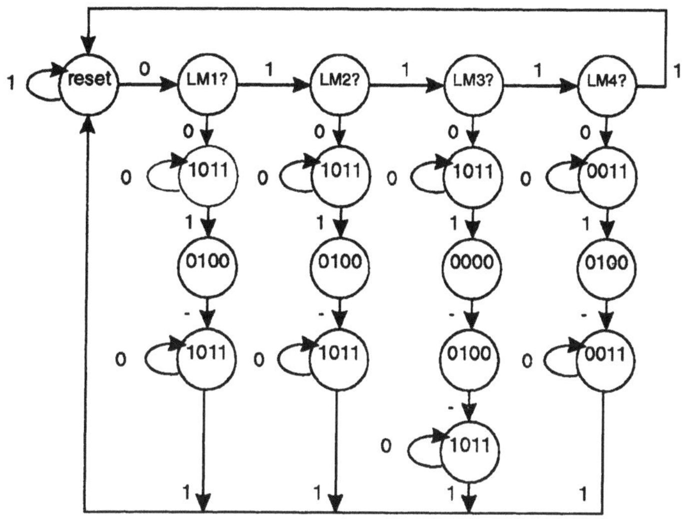

Figure 6.17. Example TCB input assignment.

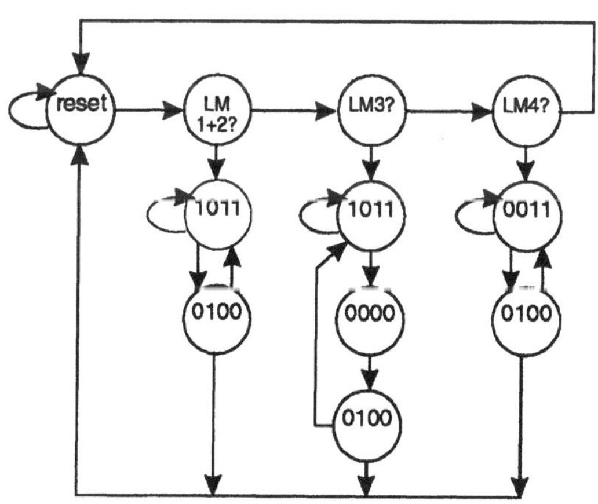

Figure 6.18. Example Test Control Block state merging.

State Assignment

The state assignment of the TCB FSM resulting from the state merging process proceeds along a binary count sequencing; see Figure 6.19. Both the state code and the output function are indicated in the state circles. A four bits state code is required. This state code is given first, separated from the output function by a '/'. For the time being, the output function of the unused states is a don't care. The first two bits of the state code represent the selection bits. The last two bits of the state code represent the test plan bits. The input assignment is also indicated to complete the TCB FSM. The unused states resulting from the test plan state code are drawn also. The last step to take is to specify the unused states also. This can be done in either of the four methods as indicated in the previous section. A first choice is to map them onto an equivalent state in the corresponding leaf-macro state diagram.

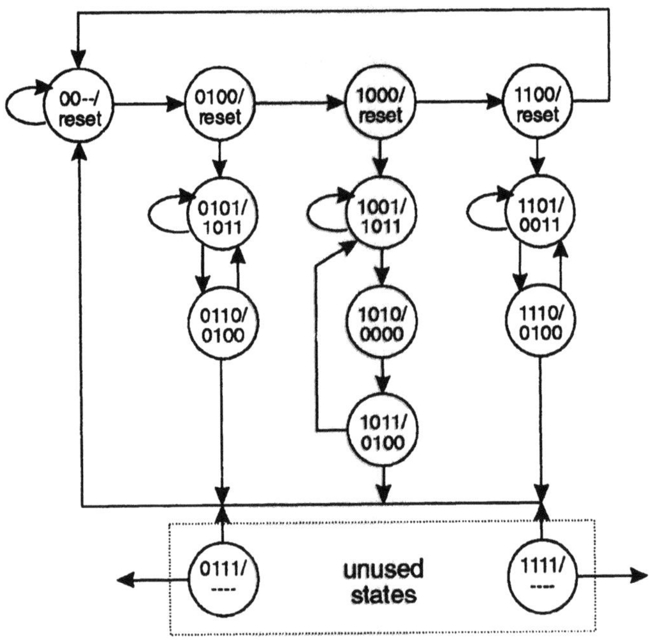

Figure 6.19. Example Test Control Block state assignment.

6.8 Distributed Test Control

One of the characteristics of modern IC design styles is that they make use of 'pre-cooked' modules. Existing design entities, such as microprocessor cores, memories, I/O logic, are embedded as black boxes in new designs. This pre-cooked approach has its consequences for the design of the Test Control Block.

Consider a design, consisting of a number of leaf-macros, where each leaf macro has a generated test plan. When this design is embedded in a larger design, then the task is to continue the process of leaf-macro accessibility, i.e., be able to test the leaf-macros from the newly created device pins. The already generated test plans become the new initial test plans and the test plan generation process continues until the boundary of the new device has been reached. The difficulty arises when the pre-cooked design contains a Test Control Block. This means that the conditional part of the leaf-macro test plan is already implemented in a piece of hardware which takes care of the local generation of test control signals. There are two ways to handle this problem.

* Leave the existing Test Control Block as it is and consider the TCB MTM input signal as a test control signal with a prescribed sequence of signal values assigned to it.
* Delete the Test Control Block from the pre-cooked module and generate an overall Test Control Block using the methods as described in the previous section.

From the point-of-view of hierarchical design, the first option should be preferred. However, what are the consequences of applying a multi-layer, distributed, test control structure.

Problem Analysis

First, we define the *scope* of a test control signal and the *scope* of a Test Control Block.

Definition 6.3.
Consider a circuit C. Let TC be the set of test control signals controlling the operation of the data flow in C and let TB be the set of Test Control Blocks in C. C consists of a hierarchical set M of macros, where C is the top level macro.

The *scope of a test control signal* tc $\in$ TC, tc-scope : TC $\rightarrow$ M, is defined as follows.

tc-scope(tc) = $\cup$ {m $\in$ M | tc controls the operation of test data flow in m}.

Identically, we define the *scope of a Test Control Block* tcb $\in$ TB, tcb-scope : TB $\rightarrow$ M,

tcb-scope(tcb) = $\cup_{tc}$ tc-scope(tc), where tc is an output signal of tcb.

In our Test Control Block analysis, we only considered a single Test Control Block and test control signals with device scope. A multi-layer test control structure involves a Test Control Block and test control signals having a non-device level scope.

Given is a circuit C having an embedded macro M. Macro M is taken in as a black-box from another design and contains a Test Control Block TCB1 with scope M; see Figure 6.20.

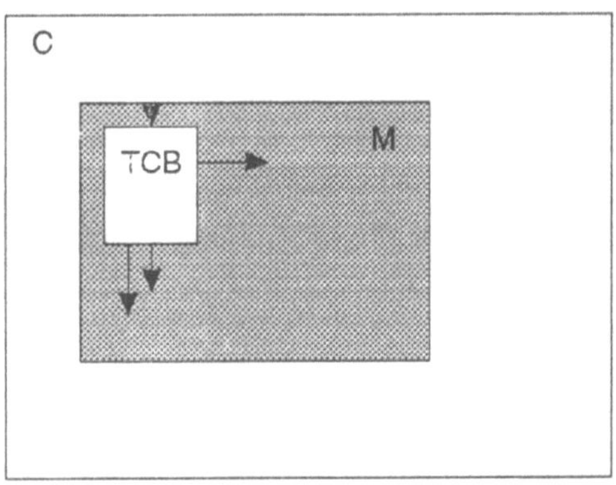

Figure 6.20. An embedded macro M having a Test Control Block.

This Test Control Block is not designed with the purpose to provide test control value sequences to control the operation of the test data path from the device ports of C to the leaf-macros in M and vice versa. Consequently, there is a problem when one or more test control signals are needed to control the operation of the data path from C to M. Basically, we either take the approach to leave the Test Control Block as it is or we remove TCB1 and generate a new Test Control Block with scope C. The latter approach is possible using the techniques described in this chapter. This is also the approach applied in all IC designs we have treated with our Macro Test techniques; see [Bouwman92a, Beenker94].

We analyze the other approach in more detail.

Leaving TCB1 creates two problems. The first problem is to control the MTM port of TCB1 from the device pins. The second problem is to provide the test control signals with scope C to control the test data paths outside M.

Controlling the MTM input port of TCB1 from the device pins is not difficult and described in [Beenker90]. For example, this MTM port can be directly connected to a device pin. There are two approaches to provide the test control signals with device scope.

1. Expand the scope of (some of) the output signals of TCB1 to C such that these signals control the operation of the test data flow from C to M, and vice versa.
2. Introduce a set of new test control signals, or a Test Control Block, with scope C, controlling the operation of the test data flow from C to M, and M to C.

The main problem is the pre-cooked TCB1 functionality. The TCB1 output function can only provide the implemented value sequences. Since these sequences were implemented with only the details of macro M in mind, they are not necessarily correct for controlling the test data path outside M. Approach 1 is only possible when the pre-cooked functionality satisfies global requirements, which means that standard test methods and access mechanisms, such as scan chains, can be supported. This standardization is in contradiction with the flexibility of the Macro Test approach.

Approach 2, which involves a multi-layer test control structure, is feasible when a clear distinction can be made between the operation of the test data path from the device pins and M and within M. In other words, TCB1 can make local adaptions to a globally defined and controlled test data path. As such, TCB1 with scope M acts as a slave controller for a master controller with scope C. Examples of such a multi-layer test control structures are described in section 6.3.

An example of a multi-layer test control application is Built-In Self-Test. A leaf-macro M with a Built-In Self-Test capability isolates itself during test application from its environment. In this case, the test control signals required for initializing, starting, and observing the Built-In Self-Test have circuit scope. The test control sequences required for executing the test protocol have scope M. An example is the SRAM Built-In Self-Test, having a local self-test controller, described in Chapter 4.

7 Exploiting Parallelism in Leaf-Macro Access

7.1 Introduction

The Macro Test concept provides a route from IC design to a set of leaf-macro test specifications. The major steps to follow are test plan generation, test pattern generation, and test pattern insertion; see Chapter 3. The test specifications contain the information of where to apply and observe given data at a specified time frame. A complete device test specification is produced by merging all leaf-macro test specifications. This process is called test assembly; see Chapter 3.

An easy test assembly procedure is a concatenation, i.e., a sequential execution of the various leaf-macro test specifications. However, concatenation may lead to a long device test time. Especially, if reloads of the tester memory are required, the testing costs may easily exceed an acceptable level [Basset90]. According to our definition of testability, as stated in Chapter 1, this results in a non-testable device.

Various approaches are possible to keep the resulting test time within reasonable limits. A scan chain routing strategy, minimizing the number of shift cycles, is described in Chapter 5 and in [Oostdijk91, Narayanan92]. Built-In Self-Test is a technique where most of the data manipulation, such as generation and observation, is done on the device itself. Scheduling techniques for various Built-In Self-Testable leaf-macros are described in [Kime82, Abadir86, Craig88, Jone89]. A disadvantage of applying Built-In Self-Test is the amount of silicon real estate used for the test pattern generators, self-test controllers, and response analyzers. The trade-off between silicon real estate and test time is probably worth the effort for large embedded memories; see Chapter 4. In other cases, Built-In Self-Test is more difficult to justify.

In this chapter, we consider test time reduction techniques which fit in the Macro Test concept and which not necessarily lead to an increase of silicon real estate for testable design. Instead of executing sequentially each leaf-macro test specification we consider *parallelism in leaf-macro access* [Bouwmeestr93, Beenker94]. It is shown that significant reductions in test time are possible in comparison with techniques where macros are tested sequentially.

7.2 Levels of Parallelism

The clue to application of parallelism in leaf-macro access can be found in the *access paths* to and from leaf-macros and the *data assigned* to these access paths. Intuitively one could say that if access paths of two leaf-macros are disjoint or if they are shared in a compatible way then these two leaf-macros can be tested in parallel. One could consider the parallelism search to be a '*resource allocation problem*'. Parallelism is possible as long as 'resources' are exclusively used or used in a compatible way. What is needed is a resource and resource compatibility definition. This is done in the next section. For the time being we consider a resource to be a design entity used for test data transport to and from a leaf-macro. We add attributes to resources, such as the clock cycle at which it is in use, and the data value assigned to it. In this section, we first explore the various levels of parallelism and we present some easy to understand examples to explain the concepts. In the following section, we provide a definition of resources and a definition of the various parallelism possibilities.

Let us first define some terminology.

Execution of the actions described in a leaf-macro generated test plan results in an execution of the leaf-macro initial test plan from the device pins. Consequently, we may state that the access path to and from a leaf-macro is inherently described in the leaf-macro generated test plan. This test plan is repeated for every leaf-macro test pattern. The execution of a generated test plan for a single pattern is called a *test plan execution.* A test plan execution is performed as a sequence of *test plan steps.* In every test plan step, test control values and transfer condition values are specified which must be applied to a set of device pins at a given clock cycle. Also, the test plan step specifies device input pins and device output pins with a *reference* to the values that must be applied to or observed from that pin. The moment the references are replaced with their actual value we obtain the *test specification step.*

Note that this implies three levels of parallelism.

Level 1. The access paths of two leaf-macros are disjoint. The access path of a leaf-macro is uniquely defined in the generated test plan as a protocol which has to be repeated for every test pattern. Hence, this level of parallelism is called *test plan parallelism.*

Level 2. It does not harm when the leaf-macros access paths have resources in common as long as they are used at different clock cycles. The usage of a resource at a certain clock cycle is described in a test plan step. Hence, this level of parallelism is called *test plan step parallelism.*

Level 3. The third level of parallelism is where leaf-macro access paths have resources in common during a number of clock cycles. Still, in this case there is a parallel access possibility. This requires a distinction to be made between the *input access path* from the device input pins to the leaf-macro input ports and the *output access path* from the leaf-macro output ports to the device output pins. Input access paths may be shared during several clock cycles when the respective data is compatible. Output access paths may never be shared at common clock cycles. Detailed resource usage is indicated in the leaf-macro test specification step. Hence, this level of parallelism is called *test specification step parallelism.*

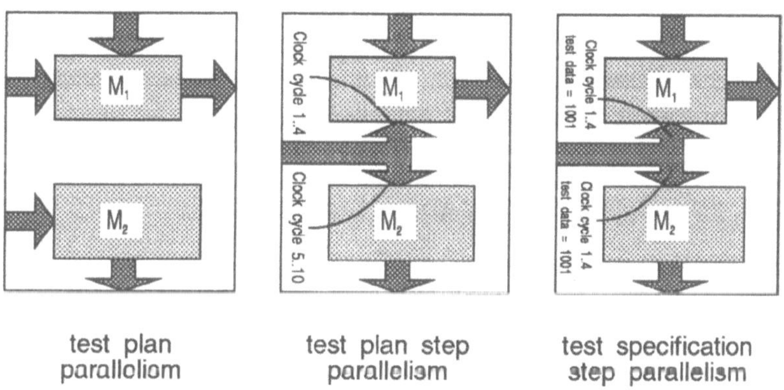

Figure 7.1. Three possible levels of parallelism in leaf-macro access.

The three levels of parallelism are symbolically indicated in Figure 7.1. The output access paths play a specific role in the third level of parallelism. Within a test specification step the *expected* response is indicated. In presence of a defect, the *actual* response might differ from the expected response. Hence, output access paths may never be shared at common clock cycles. Further, note that parallelism is not only applied between leaf-macros. A distinction between input access paths and output access paths can also be made for a singe leaf-macro. Following an identical reasoning for parallelism between leaf-macros, it follows that test plan execution of a leaf-macro test pattern input section may safely take place while performing test plan execution of the previous test pattern response section. The internal parallelism application is called *pipelining* between two consecutive test plan executions [Abadir86]. An often used example of pipelining is scanning input data into a scan chain while scanning response data out.

As stated before, resources are defined by the sequence of actions to be performed on the device input and output pins. Following this statement, we base our theoretical parallelism analysis on the assumption that a parallelism investigation can be translated to an investigation of the set of actions on device pins. However, to make things clear in this introductory section, we make the access paths explicitly visible. To explain the parallelism concepts, we present an example of each level of parallelism. Next, we develop a baseline theory on parallelism, which considers the device pins with the assigned set of actions on them, as resource.

Note that there is a dependency between the parallelism possibilities and test plan generation. Resource choices are made during the test plan generation process. The choices may hamper parallelism. The dependencies between parallelism and other Macro Test activities are described in Section 7.5.

Example of no Parallelism, Sequential Test Assembly

Consider the example circuit in Figure 7.2. It consists of two combinational leaf-macros M_1 and M_2. In this design, M_1 and M_2 are accessible via serial scan registers. The number of test patterns required to test M_1 equals n_1 and to test M_2 equals n_2, whereby $n_2 > n_1$. The modes of the four scan registers sr_1, sr_2, sr_3, and sr_4 are controlled by the test control signals tc_1, tc_2, tc_3, and tc_4, respectively.

The value assignment to the test control signals for the scan mode is specified by S, and the normal mode by N. Don't care values are specified by '-'. A scan chain routing algorithm for a single scan configuration, reducing the number of shift cycles, is given in Chapter 5. This algorithm results in the scan register ordering given in Figure 7.2. The set of resources for both M_1 and M_2 consists of the four test control signals with their assigned value, and the scan chain composed of the four scan registers.

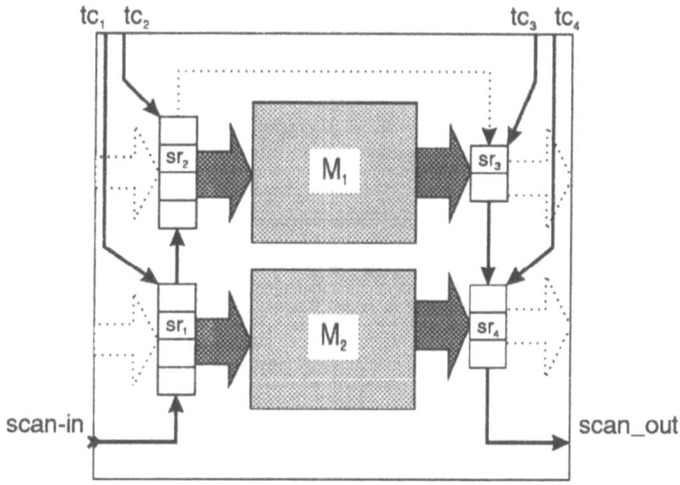

Figure 7.2. Example circuit.

Generated test plans for M_1 and M_2 are symbolically specified in Tables 7.1 and 7.2. In these test plans, $patM_1$ and $patM_2$ represent the test pattern sets that need to be applied to M_1 and M_2, respectively.

	test control				input data	response data
clock cycle	tc_1	tc_2	tc_3	tc_4	scan_in	scan_out
1..4	S	-	-	-	$patM_1[1..4]$	-
5..7	S	S	-	-	-	-
8	-	S	-	-	-	-
9	-	-	N	-	-	-
10	-	-	S	S	-	-
11	-	-	-	S	-	-
12..13	-	-	-	S	-	$patM_1[5..6]$

Table 7.1. Generated test plan for M_1.

	test control				input data	response data
clock cycle	tc_1	tc_2	tc_3	tc_4	scan_in	scan_out
1..4	S	-	-	-	$patM_2[1\text{-}4]$	-
5	-	-	-	N	-	$patM_2[5]$
6..7	-	-	-	S	-	$patM_2[6\text{-}7]$

Table 7.2. Generated test plan for M_2.

A test pattern consists of two parts. A first part specifies the bits that must be applied to the leaf-macro input ports, the second part specifies the bits that must be observed. For example, four bits must be applied to M_1, and two bits must be observed. $patM_1[1..4]$ gives a reference to the bits that must be applied, $patM_1[5..6]$ gives a reference to the bits that must be observed *as a result of the specified clock pulse.* During which tester cycle the actual response is observed is determined in the final test program where the observe window is specified. This observe window may either be at the end of the current clock cycle or before the new data transition in the following clock cycle.

A sequential test assembly procedure results in n_1 patterns times 13 cycles + n_2 patterns times 7 cycles; see Figure 7.3.

cycle	1	2	3	4	5	6	7	8	9	10	11	12	13	...	+1	+2	+3	+4	+5	+6	+7
tc_1	S	S	S	S	S	S	S								S	S	S	S			
tc_2					S	S	S	S													
tc_3									N	S											
tc_4										S	S	S	S						N	S	S
scan_in	D	D	D	D											D	D	D	D			
scan_out												D	D						D	D	D

D = test data
S= scan
N = normal

test plan M_1 repeat n_1 times

test plan M_2 repeat n_2 times

Figure 7.3. Sequential test assembly.

Example of Test Plan Parallelism

The M_1 and M_2 resource sets have scan registers sr_1 and sr_4, and test control signals tc_1, tc_2, tc_3, and tc_4 in common. Hence, there is no test plan parallelism possibility. However, test plan parallelism can be enforced by some design modifications. If the scan register routing is done in two scan chains instead of one then test plan parallelism is possible; see Figure 7.4.

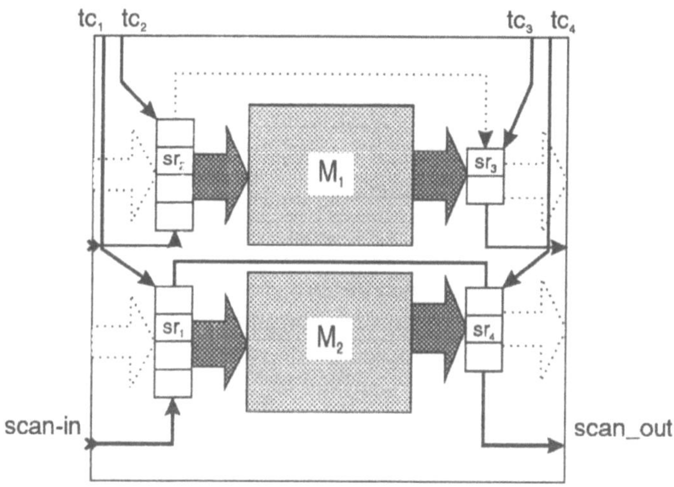

Figure 7.4. Modified example circuit allowing test plan parallelism.

Hence, a trade-off between a design modification, with possibly extra device pins, and a parallelism in leaf-macro access can be made. The device test specification resulting from test plan parallelism for M_1 and M_2 in the design of Figure 7.4 is given in Table 7.3. This device test specification takes n_2 times 7 cycles.

Note, that the test plan parallelism application increases the test time for M_1 with one clock cycle per test pattern. The application of a new test pattern for M_1 has to wait until the test plan execution of a test pattern for M_2 is finished.

		test control				input data		response data	
clock cycle	repeat	tc_1	tc_2	tc_3	tc_4	scan_in1	scan_in2	scan_out1	scan_out2
1..4	n_1	S	S	-	-	patM$_1$[1-4]	patM$_2$[1-4]	-	-
5		-	-	N	N	-	-	patM$_1$[5]	patM$_2$[5]
6		-	-	S	S	-	-	patM$_1$[6]	patM$_2$[6]
7		-	-	-	S	-	-	-	patM$_2$[7]
1..4	n_2-n_1	S	-	-	-	-	patM$_2$[1-4]	-	-
5		-	-	-	N	-	-	-	patM$_2$[5]
6..7		-	-	-	S	-	-	-	patM$_2$[6-7]

Table 7.3. Device test specification resulting from test plan parallelism.

Example of Test Plan Step Parallelism

Let us now return to the example given in Figure 7.2 and see whether we can apply test plan step parallelism without any design modification. Test plan step parallelism allows a sharing of various leaf-macro access paths as long as the involved design entities are not used at the same clock cycle. In the example circuit of Figure 7.2, scan registers sr_1 and sr_4 are shared in the access paths of M_1 and M_2. However, possibly at a different range of clock cycles. Further, test plan step parallelism for a *single* leaf-macro is possible if the leaf-macro input and output resources are not used at the same clock cycle. For example, leaf-macro M_1 uses the scan register sr_2 for applying data and scan register sr_3 for observing data. The sequence of values applied to the test control signals during the input data transport is compatible with the sequence of values applied to the test control signals during the output data transport. Hence, the input and output data shifting process data may take place in parallel. A similar argument holds for M_2. Note that the requirement on compatibility of test control signals is a somewhat hidden form of test specification step parallelism. We come back on this issue in the next section.

A test specification for M_1 according to the pipelining protocol is given in Table 7.4.

		test control				input data	response data
clock cycles	repeat	tc_1	tc_2	tc_3	tc_4	scan_in	scan_out
1..4	1	S	-	-	-	$patM_1[1\text{-}4]$	-
5..7		S	S	-	-	-	-
8		-	S	-	-	-	-
9	n_1-1	-	-	N	-	-	-
10		S	-	S	S	$patM_1[1]$	-
11		S	-	-	S	$patM_1[2]$	-
12..13		S	-	-	S	$patM_1[3\text{-}4]$	$patM_1[5\text{-}6]$
14..17		S	S	-	-	-	-
18		-	S	-	-	-	-
1	1	-	-	N	-	-	-
2		-	-	S	S	-	-
3..4		-	-	-	S	-	$patM_1[5\text{-}6]$

Table 7.4. Pipelined test specification for M_1.

The next step in exploiting the test plan step parallelism is merging the M_1 and M_2 test specifications into a device test specification. This is done by 'overlaying' the generated test plan of M_2 with the generated test plan of M_1 taking into account a number of conditions. First, common scan register resources for M_1 and M_2 may not be in use at the same clock cycle. Second, the test control value sequences required for M_1 and M_2 must be compatible such that the shift-in/apply/shift-out cycle can be applied for M_1 and M_2 simultaneously. This overlay process is specified in the following section.

After executing this parallel scheme n_1 times, all n_1 test patterns for M_1 have been applied including n_1 test patterns for M_2. The remaining n_2-n_1 test patterns for M_2 can be applied according to a similar scheme as given in Table 7.4. Table 7.5 gives the assembled test specification for the whole device resulting from test plan step parallelism. The number of cycles required for executing this device test specification equals $4n_1 + 5n_2 + 2$.

		test control				input data	response data
clock cycles	repeat	tc_1	tc_2	tc_3	tc_4	scan_in	scan_out
1..4	1	S	-	-	-	patM$_1$[1-4]	-
5..8		S	S	-	-	patM$_2$[1-4]	-
9	n_1-1	-	-	N	N	-	patM$_2$[5]
10		S	-	S	S	patM$_1$[1]	patM$_2$[6]
11		S	-	-	S	patM$_1$[2]	patM$_2$[7]
12..13		S	-	-	S	patM$_1$[3-4]	patM$_1$[5-6]
14..17		S	S	-	-	patM$_2$[1-4]	-
1	1	-	-	N	N	-	patM$_2$[5]
2		S	-	S	S	patM$_2$[1]	patM$_2$[6]
3		S	-	-	S	patM$_2$[2]	patM$_2$[7]
4..5		S	-	-	S	patM$_2$[3-4]	patM$_1$[5-6]
1	n_2-n_1-1	-	-	-	N	-	patM$_2$[5]
2..3		S	-	-	S	patM$_2$[1-2]	patM$_2$[6-7]
4..5		S	-	-	-	patM$_2$[3-4]	-
1	1	-	-	-	N	-	patM$_2$[5]
2..3		-	-	-	S	-	patM$_2$[6-7]

Table 7.5. Device test specification resulting from test plan step parallelism.

Example of Test Specification Step Parallelism

The test data contents is known during a test specification step. This provides the freedom to use this contents for multiple purposes. This can be done in two ways. First, a part of a test pattern may be saved during its application to a leaf-macro, so it can be used again as a part of a successor pattern that must be applied to the same leaf-macro. This pattern overlay process is applied in [Leenstra93b]. Second, parts of test data which are required for more than one leaf-macro may be applied to these macros simultaneously. Examples are given from both applications.

First, an example in which a part of a test pattern is saved during its application to a leaf-macro and used again as a part of a successor pattern that must be applied to the same leaf-macro. Again, consider Figure 7.2. Let $n_1 = 3$ and the test pattern set for M_1 be {1101LH, 0101HH, 0110LL}. Note from Figure 7.2 that for M_1 the applying scan register sr_2 is four scan cells wide, whereas the observing scan register sr_3 is only two scan cells wide. Thus, sr_3 can observe a new response pattern from M_1 every two clock cycles. Without test data contents knowledge it takes four cycles to fill sr_2 with a new pattern for M_1. However, if the test data content is taken into consideration, a possible overlap between the test patterns can be exploited. Assume sr_2 is filled with test data for M_1. Also, assume that sr_1, sr_2, and sr_4 contain a mode in which the contents of these scan registers is not lost during the apply mode. This mode is called the apply/hold mode (A). Now, overlap between test patterns can be exploited by shifting two instead of four times after each pattern apply cycle. The resulting M_1 test specification from this test specification step parallelism is given in Table 7.6.

Second, an example in which test data is simultaneously applied to two leaf-macros. Consider the example design in Figure 7.5. In this design, M_1 and M_2 are connected to a common scan register. A test pattern that has been scanned into this scan register can be applied to M_1 and M_2 simultaneously. The responses are observed in scan registers which are not used simultaneously by M_1 and M_2. An interesting approach to minimize the test time of this design could be to try to find the minimum test pattern set that detects faults in both M_1 and M_2 instead of generating test pattern sets for M_1 and M_2 separately. If patterns for M_1 and M_2 are generated using different test pattern generation tools, e.g., if M_1 is combinatorial logic and M_2 is a RAM, a possible method could be to generate test patterns for M_1 and to check via fault simulation which faults in M_2 are covered by this test pattern set. For M_2, test pattern generation is now only required for the remaining faults. If M_1 and M_2 are identical, considering test specification step parallelism provides an effective approach for parallel testing [Mehtani90].

	test control				input data	response data
clock cycle	tc_1	tc_2	tc_3	tc_4	scan_in	scan_out
1..4	S	-	-	-	1101	-
5..8	S	S	-	-	0110	-
9	A	A	N	-	-	-
10	S	S	S	S	-	-
11	S	S	-	S	-	-
12	A	A	N	A	-	-
13	S	S	S	S	-	L
14	-	S	-	S	-	H
15	-	-	N	A	-	-
16	-	-	S	S	-	H
17..19	-	-	-	S	-	HLL

Table 7.6. M_1 test specification resulting from test specification step parallelism.

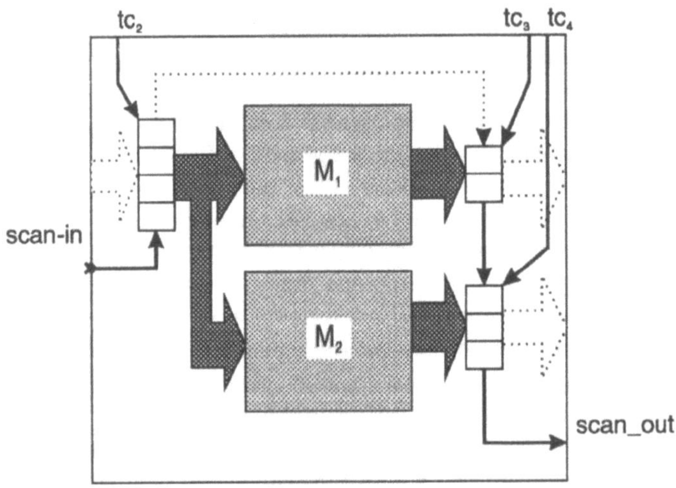

Figure 7.5. Example circuit for test specification step parallelism.

7.3 Formal Definitions of Resources, Resource Compatibility and Parallelism

As we have shown via an intuitive approach and some examples, exploiting parallelism in leaf-macro access can be considered as a resource allocation problem. We now formalize this approach. Resources are defined as device pins which may be used during a certain clock cycle for applying or observing data. Given this definition, the various levels of parallelism are expressed in terms of resources. A next step is to define various relations between resources, based on which necessary and sufficient resource conditions for the various forms of parallelism can be deduced. These conditions form the basis for an algorithmic procedure for deducing leaf-macro parallelism possibilities.

We limit our parallelism theory to forms of parallelism where a generated test plan of one leaf-macro can be executed within the execution interval of a generated test plan of another leaf-macro, see Figure 7.6. This excludes trivial forms of parallelism where parallelism can only be applied during a few cycles.

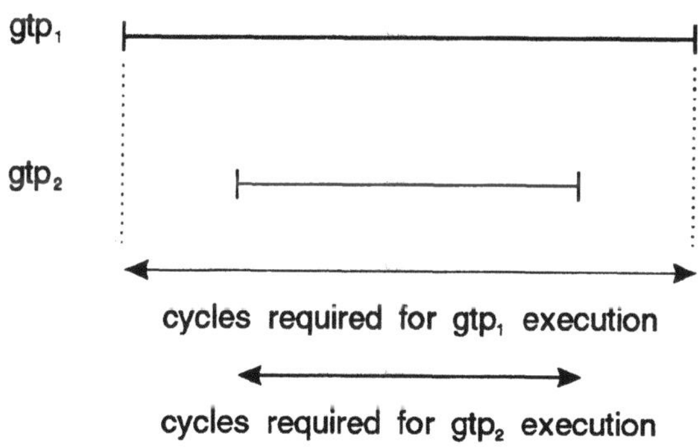

Figure 7.6. The considered form of parallelism.

We start with the definition of resources. Resources are determined during the test plan generation process. This process takes place on a representation of the device. This representation is normally a hierarchical description of the device in terms of *design entities*: library elements or groups of library elements and wires connecting various ports. During the test plan generation process, test data transport design entities are appointed, and conditional values are assigned for enabling the test data transport. Examples of such design entities are scan registers, cells having transfer properties, test control signals, and transfer condition signals. Eventually, all design entities corresponding to the leaf-macro access path are uniquely assigned in the corresponding generated test plan. This assignment is in terms of a sequence of actions to be performed on a set of device pins during given clock cycles.

After a successful test plan generation process, all leaf-macro resources are determined. Given this information, the next step is to study the possibility for parallelism in leaf-macro access. These two issues relate to each other. We have chosen for the set up to consider the generated test plans as given items. Another choice could have been to state requirements for parallel leaf-macro access as a boundary condition for test plan generation. This would have complicated the theory of parallel leaf-macro access too much and is a topic for further research.

We now have the following descriptive definition of a generated test plan resource.

A *generated test plan resource*, or in short *a resource* is a device pin on which a sequence of actions during a number of clock cycles is assigned in the generated test plan. We distinguish between *input resources*, *output resources*, and *control resources*. An input resource is used during a number of clock cycles for applying input data. An output resource is used for observing leaf-macro response data during a number of clock cycles. Control resources are used for assigning values during given clock cycles to test control signals and transfer condition signals.

This resource definition comprises one basic assumption about the test plan generation process. When two generated test plans do not make use of common device pins then the two leaf-macro access paths do not share any design entity. Also, if two generated test plans do share device pins, however, at different clock cycles, then there are no design entities shared in the respective access paths at identical clock cycles. In other words, sharing in access paths is directly translated into sharing of device pins, and vice versa. This assumption is topic for further research.

The following difference exists between the various resources. Control resources have an assigned value at a certain clock cycle already in the generated test plan. Input resources have an associated value during a clock cycle which is instantiated in the test specification resulting from the test pattern insertion process. The output resources have an assigned expected value during a clock cycle which is also be instantiated as a result of the test pattern insertion process.

Let **GTP** be the given set of generated test plans,

$$\mathbf{GTP} = \{ \text{ gtp} \mid \text{gtp is a generated test plan}\},$$

and let **R** be the set of available resources,

$$\mathbf{R} = \{ \text{ r} \mid \text{r is a resource } \}.$$

For each gtp $\in$ **GTP** we define the input resource set $R^{in}(gtp)$, the output resource set $R^{out}(gtp)$ and the control resource set $R^{con}(gtp)$.

$R^{in}(gtp) =$ { r $\in$ **R** | r is a device pin assigned in gtp and used for applying input data},

$R^{out}(gtp) =$ { r $\in$ **R** | r is a device pin assigned in gtp and used for observing response data},

$R^{con}(gtp) =$ { r $\in$ **R** | r is a device pin assigned in gtp and used for applying values to test control signals and transfer condition signals}.

The resource set R(gtp) of generated test plan gtp is the union of $R^{in}(gtp)$, $R^{out}(gtp)$ and $R^{con}(gtp)$.

$$R(gtp) = R^{in}(gtp) \cup R^{out}(gtp) \cup R^{con}(gtp).$$

Note that the sets $R^{in}(gtp)$ and $R^{con}(gtp)$ are not necessarily disjoint. The same holds for $R^{in}(gtp)$ and $R^{out}(gtp)$.

Example

Consider the example circuit in Figure 7.7. Two leaf-macros are indicated including a design entity with a transfer property. Both leaf-macros are accessible via scan, where one input of leaf-macro M_2 needs a transfer information. The transfer condition signal is supplied as output of scan cell sc_4 and is transported via scan_in, sc_1, sc_2 and sc_3. The various resource sets are as follows.

$R^{in}(gtp_1) = \{scan\_in\}$, $R^{con}(gtp_1) = \{tc_1\}$, $R^{out}(gtp_1) = \{scan\_out\}$,

$R^{in}(gtp_2) = \{scan\_in, A\}$, $R^{con}(gtp_2) = \{scan\_in, tc_1\}$, $R^{out}(gtp_2) = \{scan\_out\}$.

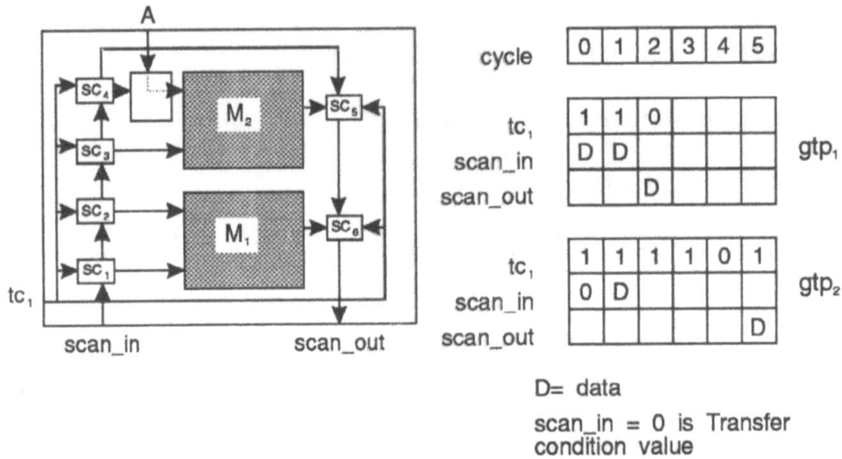

Figure 7.7. Example circuit with indicated generated test plans gtp_1 and gtp_2.

Having defined the notion of resources we can now express the various parallelism levels in terms of resources.

Definition 7.1: Test plan, test plan step and test specification step parallelism

Consider two leaf-macros M_1 and M_2 with their corresponding generated test plans gtp_1 and gtp_2. The generated test plans are said to be *executed in parallel* if *within* the required number of cycles for a test plan execution of the one macro a test plan execution of the other macro can take place; see Figure 7.6. This parallelism is called

* *Test plan parallelism* if and only if the generated test plans have no input/output resources in common and for common control resources holds that they either are used at different cycles or used at identical cycles with a compatible value.

* *Test plan step parallelism* if and only if the generated test plans are either test plan parallel or if common resources are shared during different clock cycles and if common control resources at common cycles have compatible values.
* *Test specification step parallelism* if and only if the generated test plans are test plan step parallel or if common input and control resources are shared at common cycles with compatible values and common output resources are used at different clock cycles.

From this definition it follows that two resource attributes are of importance. The clock cycles at which a resource is in use and the value assigned to the resource during these clock cycles.

Consider a generated test plan gtp $\in$ **GTP** with its associated input, output, and control resource set. In order to be able to compare cycles we assume for the time being that the actions defined in the generated test plan always start at the fixed cycle 0. We define two functions on these resource sets. The first function defines the clock cycles at which a certain resource is in use. The second function defines the value assigned to a resource during a certain cycle.

Let **N** be the set of natural numbers, **N** = {0, 1, 2, ...}. Let r $\in$ R(gtp), and $\wp(\mathbf{N})$ the set of all possible subsets of the natural numbers **N**, i.e., the power set of **N**.

$$\text{Time: gtp} \times R(\text{gtp}) \rightarrow \wp(\mathbf{N}),$$
$$\text{Time}(\text{gtp}, r) := \{ n \in \mathbf{N} \mid r \text{ is used as resource for gtp at cycle } n \}.$$

Let r $\in$ $R^{in}(gtp) \cup R^{con}(gtp)$ and n $\in$ **N** then

$$\text{Value: gtp} \times R(\text{gtp}) \times \mathbf{N} \rightarrow \{0, 1, -\},$$
$$\text{Value}(\text{gtp}, r, n) := \text{value assigned to } r \text{ at cycle } n \text{ during execution of gtp}.$$

The Value function can only be defined on input and control resources. Output resources have an expected output value. The actual output value may differ from the expected output value.

Example

Again, consider the example circuit in Figure 7.7. Assuming that we apply D = 1, A = 1, and transfer condition value = 0, we have the following definition of the Time and Value function.

Time(gtp_1, scan_in} = {0,1},	Value(gtp_1, scan_in, 0) = 1,
	Value(gtp_1, scan_in, 1) = 1,
Time(gtp_1, tc_1} = {0,1,2},	Value(gtp_1, tc_1, 0) = 1,
	Value(gtp_1, tc_1, 1} = 1,
	Value(gtp_1, tc_1, 2) = 0,
Time(gtp_1, scan_out} = {2},	
Time(gtp_2, scan_in) = {0,1},	Value(gtp_2, scan_in, 0} = 0,
	Value(gtp_2, scan_in, 1} = 1,
Time(gtp_2, tc_1} = {0,1,2,3,4,5},	Value{gtp_2, tc_1, 0} = 1,
	Value(gtp_2, tc_1, 1} = 1,
	Value(gtp_2, tc_1, 2} = 1,
	Value(gtp_2, tc_1, 3) = 1,
	Value(gtp_2, tc_1, 4} = 0,
	Value(gtp_2, tc_1, 5} = 1,
Time(gtp_2, A} = {4},	Value(gtp_2, A, 4) = 1,
Time(gtp_2, scan_out} = {5}.	

If an input or control resource is not indicated at a certain clock cycle it means it has a value '-', i.e., don't care.

Now, we can reason about resources with assigned attributes such as cycle time and assigned value. As follows from the definition of parallelism, parallelism between two generated test plans can take place if they have different resources, or if they share resources at different time steps, or if they share input/control resources during a number of time steps with a compatible value, with the additional requirements that output resources may not be shared at common time steps. The functions Time and Value are defined for the purpose of reasoning about resources at time steps with an assigned value. The next step to take is to define when two resources are *compatible*. Compatible in the sense that they are either different, or used during different clock cycles or used during common clock cycles with a compatible value. Having defined compatible resources we can derive the conditions for the various forms of parallelism.

We assumed that every generated test plan starts at cycle 0. We now relax this assumption and allow any cycle as long as the execution of the one test plan takes place in parallel with the execution of another test plan; see Figure 7.6. Hence, we may shift the cycles of a generated test plan forward in time with a certain positive integer number.

We start with the control resources. Since control resources have an assigned value, they are the most demanding resources for parallelism. In all forms of parallelism we have to consider the Value function of the control resources. In fact, this is considering test specification step parallelism for control resources already at test plan and test plan step parallelism level. It is in the next levels of parallelism that the Value function of input resources is taken into account as well.

Control values enable the test data flow. This means that for testing two leaf-macros in parallel, the test data flow has to be enabled in parallel. This can either be done if the involved control resources are [1] disjoint, [2] identical but used at different cycles, or [3] if mutual control resources have compatible states at mutual cycles. If a control resource is shared between two generated test plans, however, at different cycles (case 2) we say that the control resource is *time-disjoint.* In case 3) we say that one control resource *covers* the other control resource. This time-disjoint and cover property is expressed in the following relations. Remind that we may shift the start cycle of a generated test plan forwards in time.

Definition 7.2: Time-disjoint property of two resources

Let $r_1 \in R(gtp_1)$ and $r_2 \in R(gtp_2)$. Let N1 be the number of test plan steps in gtp_1 and N2 be the number of test plan steps in gtp_2, with N1 > N2. We say that resource r_1 and resource r_2 are *time-disjoint*

$$r_1 \varnothing_{td} r_2 \quad \Leftrightarrow \quad r_1 = r_2 \wedge \exists_{k \in \mathbf{N},\, 0 \le k \le N1-N2} \left[(Time(gtp_2, r_1) + k) \cap Time(gtp_1, r_2) = \varnothing \right],$$

where $Time(gtp_2,r_1) + k = \{n + k \mid n \in Time(gtp_2,r_1)\}$.

We call k the *k-factor* of the two resources.

The time-disjoint property is used for control resources in case of test plan parallelism and for all resources for the other forms of parallelism.

Next, we define in an similar way the cover property of two input or control resources. Parallelism is still possible if an input or control resource is shared between two input or control resource sets with a compatible value. This is called the *cover property.*

Definition 7.3: Cover property of two resources

Let $r_1 \in R(gpt_1)\backslash R^{out}(gtp_1)$ and $r_2 \in R(gtp_2)\backslash R^{out}(gtp_2)$. Let N1 be the number of test plan steps in gtp_1 and N2 be the number of test plan steps in gtp_2, with N1 > N2. We say that resource r_1 *covers* resource r_2

$$r_1 \sqsubset r_2 \Leftrightarrow \quad r_1 = r_2 \wedge$$
$$\exists_{k \in \mathbf{N},\, 0 \le k \le N1-N2} \forall_{0 \le n < N2} \left[\text{Value}(gtp_1, r_1, n+k) \sim \text{Value}(gtp_2, r_2, n) \right],$$

where two values are compatible if and only if we can create a new value which is covering the functionality of both values; see Chapter 6. Again, we have taken the shift possibility by k cycles into account.

The cover property of two control resources is considered during all forms of parallelism. The cover property of two input resources is only considered for test specification step parallelism. To explain this definition we give an example of the cover property.

Example of Cover Property of Two Control Resources.

Consider the example circuit in Figure 7.7. The device pin tc_1 is a control resource of both gtp_1 and gtp_2. N1 = 3, N2 = 6. It is easy to see that $\text{Value}(gtp_2, tc_1, n+2) \sim \text{Value}(gtp_1, tc_1, n)$ for each $n \in \{0, 1, 2\}$.

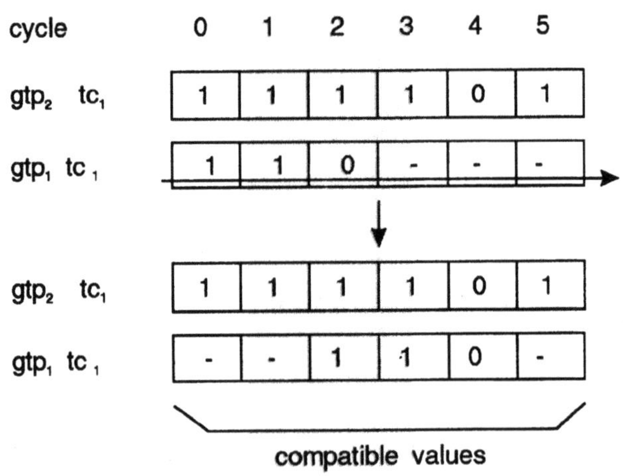

Figure 7.8. Cover property of mutual control resources.

Hence, control resource $tc_1 \in R^{con}(gtp_2)$ covers control resource $tc_1 \in R^{con}(gtp_1)$; see Figure 7.8.

Having defined the required properties of resources we can now define test plan compatibility of two generated test plans. Two generated test plans are *test plan compatible* if and only if their resource sets have no common input/output resources or if they have common control resources then they have to be time-disjoint or the one control resource set covers the other control resource set. However, if we shift the cycles of one generated test plan forwards in time by k steps, this is done uniquely for all actions described in this generated test plan. Hence, the k-factor can only be defined once. This is expressed in the following definition.

Definition 7.4: Test plan compatibility of two generated test plans

Generated test plans gtp_1 and gtp_2 with resource sets $R(gtp_1)$ and $R(gtp_2)$ are *test plan compatible*

$gtp_1 \sim_{tp} gtp_2 \Leftrightarrow$

$$(R^{in}(gtp_1) \cup R^{out}(gtp_1)) \cap (R^{in}(gtp_2) \cup R^{out}(gtp_1)) = \emptyset \wedge$$

$$\exists! \text{k-factor } \forall r_1, r_2 \in R^{con}(gtp_1) \cap R^{con}(gtp_2) \; [r_1 \emptyset_{td} r_2 \;\vee r_1 \sqsubset r_2 \;\vee r_2 \sqsubset r_1].$$

Note that if there is a unique k-factor such that $r_1 \sqsubset r_2 \wedge r_2 \sqsubset r_3$, then $r_1 \sqsubset r_3$ holds for the same k-factor.

Having a formal definition of test plan compatibility we can now express the informal definition of test plan parallelism in a formal way formal way.

Definition 7.5: Test plan parallelism

Let $gtp_1 \in$ **GTP** and $gtp_2 \in$ **GTP**. *Test plan parallelism* of gtp_1 and gtp_2 exists if and only if $gtp_1 \sim_{tp} gtp_2$.

The next level of compatibility to be defined is test plan step compatibility. Test plan step parallelism of two generated test plans is possible if common input/output resources are used during different cycles. This is expressed by the time-disjoint property of two resources. Further, common control resources are either time-disjoint or the one covers the other. The latter is a similar requirement as for test plan compatibility. In a similar manner as for test plan compatibility we pose the test plan step compatibility definition of two generated test plans.

Definition 7.6: Test plan step compatibility of two generated test plans

The two generated test plans gpt_1 and gtp_2 with corresponding resource sets $R(gtp_1)$ and $R(gtp_2)$ are test plan step compatible

$$gtp_1 \sim_{tps} gtp_2 \Leftrightarrow$$
$$R(gtp_1) \cap R(gtp_2) = \emptyset \vee$$

$$\exists!\text{k-factor } [\ \forall r_1, r_2 \in (R^{in}(gtp_1) \cup R^{out}(gtp_1)) \cap (R^{in}(gtp_2) \cup R^{out}(gtp_2))$$
$$[r_1 \emptyset_{td} r_2\ ] \wedge$$

$$\forall r_1, r_2 \in R^{con}(gtp_1) \cap R^{con}(gtp_2)\ [r_1 \emptyset_{td} r_2 \vee r_1 \sqsubset r_2 \vee r_2 \sqsubset r_1\ ]\ ].$$

In a similar manner as for test plan parallelism we can now formalize the definition of test plan step parallelism.

Definition 7.7: Test plan step parallelism

Let $gtp_1 \in \mathbf{GTP}$ and $gtp_2 \in \mathbf{GTP}$. *Test plan step parallelism* of gtp_1 and gtp_2 exists if and only if $gtp_1 \sim_{tps} gtp_2$.

Example of test plan step parallelism.

Consider the example circuit in Figure 7.7. Generated test plans gtp_1 and gtp_2 share the input resource scan_in, output resource scan_out and control resource tc_1. Control resource $tc_1 \in R^{con}(gtp_2)$ covers $tc_1 \in R^{con}(gtp_1)$ if k-factor = 2. For the same k-factor, it holds that the input resource scan_in is used at cycles 0,1 for M_2 and cycles 2,3 for M_1. Similarly, the common output resource scan_out is used at cycle 4 for M_1 and cycle 5 for M_2. The resulting parallel test plans are indicated in Table 7.7.

	test control	input data		response data
clock cycles	tc_1	scan_in	A	scan_out
0	1	trans.cond.		-
1	1	$patM_2[2]$		-
2..3	1	$patM_1[1\text{-}2]$		-
4	0	-	$patM_2[1]$	$patM_1[3]$
5	1	-		$patM_2[3]$

Table 7.7. Test plan step parallelism for M_1 and M_2.

As a final step, we define test specification step compatibility of two generated test plans.

Definition 7.8: Test specification step compatibility of two resource sets

The two generated test plans gtp_1 and gtp_2 with corresponding resource sets $R(gtp_1)$ and $R(gtp_2)$ are *test specification step compatible*

$gtp_1 \sim_{tss} gtp_2 \Leftrightarrow$

$$R(gtp1) \cap R(gtp2) = \varnothing \vee$$

$$\exists!\text{k-factor}$$

$$[\ \forall\ r_1,r_2 \in R(gtp_1) \cap R(gtp_2)\ [\ r_1\ \varnothing_{td}\ r_2\ ] \vee$$

$$\{\ \forall\ r_1,r_2 \in (R^{in}(gtp_1) \cup R^{con}(gtp_1)) \cap (R^{in}(gtp_2) \cup R^{con}(gtp_2))$$
$$[r_1 \sqsubset r_2 \vee r_2 \sqsubset r_1\ ] \wedge$$

$$\forall\ r_1,r_2 \in (R^{out}(gtp_1) \cup R^{out}(gtp_2))\ [\ r_1\ \varnothing_{td}\ r_2\ ]$$

$$\}$$
$$]$$

Test specification step compatibility is the formal framework to define test specification step parallelism.

Definition 7.9: Test specification step parallelism

Let $gtp_1 \in$ **GTP** and $gtp_2 \in$ **GTP**. *Test specification step parallelism* of gtp_1 and gtp_2 exists if and only if $gtp_1 \sim_{tss} gtp_2$.

7.4 Test Compatibility Graphs

The results of a compatibility analysis can be presented in a test compatibility graph [Kime82]. There exists a test compatibility graph for each level of parallelism. Every node of a test compatibility graph represents a resource set corresponding to a generated test plan. Depending on the kind of parallelism we are searching for, an edge between two nodes occurs if the two resource sets are compatible in the required form. Given the test compatibility graphs, the task of searching for parallelism becomes manageable. The problem still exist to find a test scheduling algorithm which provides a device test specification with a minimum number of required tester cycles.

Example of a Test Compatibility Graph

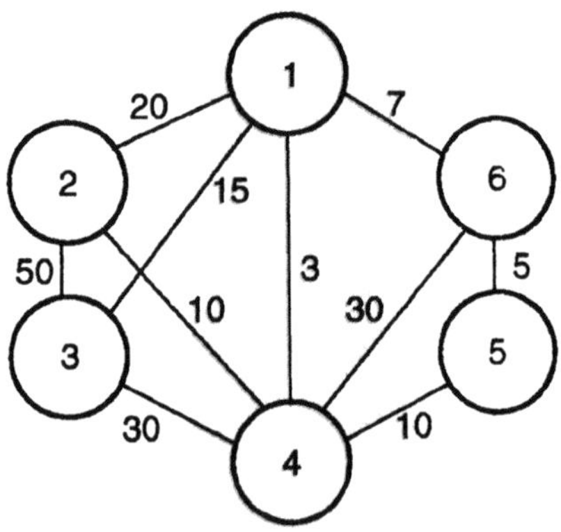

Figure 7.9. Example of a test compatibility graph for test plan compatibility.

In Figure 7.9, an example is given of a test compatibility graph for test plan compatibility. The graph has six nodes representing the six generated test plans. Test plan compatibility is expressed in terms of an arc between two nodes. Along each arc a value is given which is equal to the amount of cycles which can be saved when the two generated test plans can be executed in parallel instead of in sequence. The scheduling procedure we have taken is similar to the one as described by [Kime82, Abadir86] and is as follows.

1. Search for the two test plan compatible generated test plans sets with the highest value along its arc. In the example these two generated test plans are gtp_2 and gtp_3.
2. Compute the *maximum compatible subgraph* to which these test plan compatible generated test plans belong. A maximum compatible subgraph is a clique of a graph where *the total set of nodes* of nodes is compatible. Assume that in the example of Figure 7.9 this maximum compatible subgraph consists of gtp_1, gtp_2, gtp_3 and gtp_4. From the definition of test plan compatibility we conclude that the nodes of this clique are test plan parallel.
3. Delete the nodes which have been considered from the graph and consider the nodes not yet taken into account and start again at 1. Process until all nodes have been treated.

Implementation results of this test scheduling algorithm are provided in section 7.6 and [Bouwmeestr93].

7.5 Resource Allocation versus Test Assembly

Test assembly is the process of producing the device test specification from the set of leaf-macro test specifications. It is interesting to see how applying parallelism during test assembly influences a lot of the mentioned Macro Test activities such as test plan generation, test pattern generation, Test Control Block generation, etc. In this section we analyze the place of test assembly in the whole Macro Test process and we discuss the influence of resource allocation and resource definition on the final outcome.

Test Assembly and Test Plan Generation

During test plan generation the leaf-macro resource sets are determined. The choices made during the resource allocation are not necessarily guided by the resource compatibility requirements. A certain resource choice might destroy the parallel testing possibilities for certain leaf-macros. This consequence becomes visible after the test assembly process, unless the parallel test requirements are taken into account during test plan generation. Further, if incompatibility between test plans cannot be removed, design modifications can be made by adding testability hardware structures. Analysis of occurring resource sharing conflicts may lead to the conclusion that design modifications can further reduce test time. A trade-off between additional testability hardware and a reduced test time can now be made.

Test Assembly and Test Control Block Generation

A compatibility relation between control resources is one of the requirements for parallel testing. This certainly holds for test control signals. Parallel test requirements may pose restrictions on the value assignment to the test control resources. This value assignment is normally an action taken during Test Control Block generation. During this process, don't care values are used for an optimal implementation. There are two ways to circumvent an inability of parallel access due to non-compatible test control resources. Either, put restrictions on the Test Control Block generation by forcing as much as possible towards compatibility. Or, redo the Test Control Block generation after the test assembly process and create a loop in the activities.

Test Assembly and Scan Chain Routing

The model as set-up for scan chain routing in Chapter 5 does not take into account that there are parallel test possibilities. Hence, there is no guarantee that the existing scan chain routing is optimal with respect to test time when parallelism is considered. This brings us back to the definition of a leaf-macro as given in Section 3.1.4. The result of test assembly is a schedule of the various generated test plans with the respective test data filled in. This schedule consists of a number of parts. In each part, a set of test patterns is applied to a number of leaf-macros in parallel. We can extend the concept of a leaf-macro to a virtual leaf-macro. A virtual leaf-macro can be tested by executing its assembled test plan for every pattern in its assembled test pattern set. By considering the set of virtual leaf-macros, the scan chain routing could be done again based on the new figures of number of test patterns and usage of scan registers. Again, this creates a loop in the total set of activities.

Test Assembly and Test Pattern Generation

Another influence of test assembly is on test pattern generation. Taking contents of test data into account during application of test specification parallelism poses the question on whether with different test data equal test quality can be obtained with less test time. A set of test patterns generated with the background knowledge of testing a set of leaf-macros in parallel might be better in terms of test data volume compared to the set of test patterns generated separately for each leaf-macro without considering parallelism.

Flexibility in Choice of Resource Allocation

There is an interesting aspect in the relationship between resource allocation and test assembly which has not yet been discussed. The more compatibility exists the more possibilities there are for parallelism. Considering the definitions of compatibility one may note that the compatibility on control resources is the most demanding. A great flexibility in accessing leaf-macros may easily result in a complex sequence of incompatible control values. This in itself is a big plead for standard access mechanisms, such as scan design, with standard control value sequences.

7.6 Algorithmic Implementation and Experimental Results

Experiments have been performed with the a parallel test assembly algorithm exploiting test plan and test plan step parallelism [Bouwmeestr93]. The assembly algorithm generates a test specification for a number of macros with compatible resources for test plan and test plan step parallelism. The input for the algorithm is the set of generated test plans and the sets of test patterns for every leaf-macro.

Whenever possible, parallelism is exploited and the shift-in of new patterns is done concurrently with the shift-out of the responses of the previously applied pattern. The limitation is that scan access to every leaf-macro is assumed.

Using the resource specifications, test plan and test plan step compatibility can be determined between every pair of generated test plans and visualized in test compatibility graphs. Using the test compatibility graphs, a schedule of test plans with minimal or close to minimal test time can be determined via the algorithm described in section 7.4.

To demonstrate the validity of the parallel leaf-macro access approach we have performed experiments on a set of complex ICs. The first experiment was a digital signal processor (DSP1) IC. The DSP1 contains about 150.000 transistors and is partitioned into 18 leaf-macros. For every generated test plan, the number of leaf-macro test patterns and the maximum length of scan chains involved in testing the leaf-macro have been given in Table 7.8. Further, the compatibility classes for the control resource sets are given. This means that leaf-macros with the same control class number have compatible control resources sets. Hence, they are either time-disjoint or covering.

It turned out that for all leaf-macros the control resources were compatible between the scan-in and scan-out cycles. Hence, a pipelined protocol could be applied for all leaf-macros. The number of required tester cycles for sequentially testing these leaf-macros while applying the pipelined protocol is indicated at the bottom row of Table 7.8. All further improvements in test time are compared with this number.

The test compatibility graph for test plan step compatibility for the DSP1 is given in Figure 7.10. Based on this test compatibility graph, we have computed the test plan step parallelism possibilities. The result of this process is indicated in Table 7.9. For every assembled test plan the number of test patterns and the maximum shift cycle involved is indicated. As can be seen, the test time dropped to 466889 cycles, which means a reduction of 16%.

i	macro type	control class	$N(t_i)$	shift length
1	PLA	1	21	110
2	controller	1	80	110
3	interface	2	64	91
4	controller	2	20	94
5	interface	3	20	40
6	interface	3	80	59
7	detector	3	20	202
8	ROM	1	20	73
9	FSM	1	20	72
10	controller	1	363	72
11	FSM	1	603	55
12	ALU	3	10	192
13	ROM	3	192	146
14	multiplier	3	150	158
15	ROM	3	76	96
16	RAM	3	1714	91
17	RAM	3	1456	163
18	ACU	3	50	69
sequential test assembly, pipelined protocol				553234

Table 7.8. Leaf-macro data for the DSP1.

The following step in our experiments was to analyze the places where input and output resource conflicts occur. The results of this analysis are shown in Table 7.10. The first row of this table shows that a reduction of 50% is reached if the input or output resources are shared. Thus, for this DSP1 device, 50% is the minimum reduction that can be obtained with test plan step parallelism. The other rows of Table 7.10 indicate the result when test plan step incompatibility between various test plans is removed. The incompatibility between generated test plans 16 and 17, i.e., the test plans for the two RAMs, is removed by choosing a different test plan. This resulted in a reduction of 40% and the removal of this incompatibility did not require any design modification.

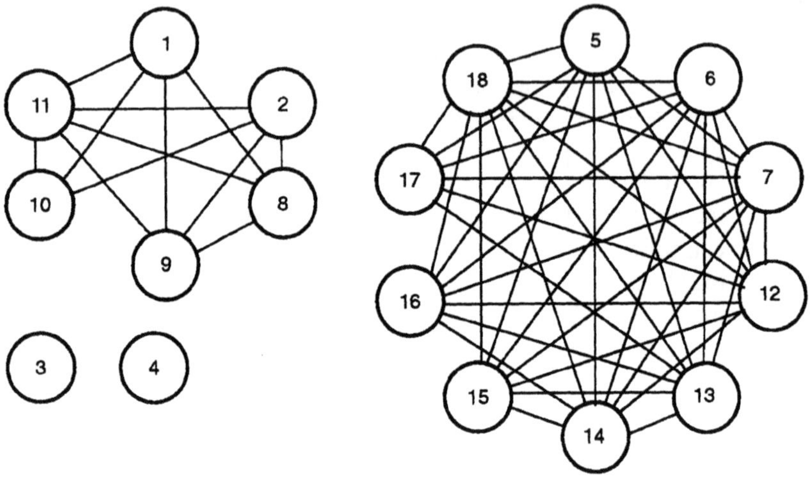

Figure 7.10. Test compatibility graph for test plan compatibility for the DSP1.

i	leaf-macro components					$N(t_i)$	shift length
1	6	7	13	17	18	20	202
2	6	13	17	18		30	163
3	6	13	17			30	163
4	5	13	17			20	163
5	13	17				92	163
6	12	17				10	192
7	17					1254	163
8	14	16				150	158
9	16					1564	91
10	2	10	11			80	110
11	1	10	11			21	110
12	10	11				262	72
13	8	9	11			20	73
14	11					220	55
15	15					76	96
16	3					64	91
17	4					20	94
Test plan step parallelism cycles							466889

Table 7.9. Test plan step parallelism results for the DSP1.

test time	reduction	removed incompatibility
276224	50%	all input/output incompatibility
332937	40%	16-17
456839	17%	14-17
459443	17%	15-16
459517	17%	15-17
465049	16%	3-4
466091	16%	1-2
466869	16%	8-10
466889	16%	5-6
466889	16%	9-10
466889	16%	12-13

Table 7.10. Test time reduction if test plan step incompatibilities between generated test plans are removed.

Application of test plan step parallelism to a second industrial device has lead to similar promising results. A general overview of results for both ICs given in Table 7.11. Again, some test plan step incompatibilities could be removed by choosing a different generated test plan. In general, removal of incompatibilities can only be done by a design modification.

IC	method	test time	reduction
DSP1	sequential	1099616	
	pipelined	553234	0%
	test plan step parallelism	466889	16%
	test plan step parallelism after removing some incompatibilities	355721	40%
	removing all input/output incompatibilities	276224	50%
Dig. Filter	sequential	1170360	
	pipelined	587076	0%
	test plan step parallelism	298894	49%
	removing all input/output incompatibilities	136932	77%

Table 7.11. Test time reduction results on two ICs.

8 Timing Aspects of CMOS VLSI Circuits

8.1 Introduction

In modern VLSI circuits timing design and timing verification have become serious design problems. In submicron digital integrated circuits the switching speed of the gates tends to be about 0.1 - 0.2 ns. Due to the increase in speed, the physical design of a circuit is no longer a low level electronic design problem. The logic designer is increasingly confronted with physical effects that have a negative influence on the robustness and noise immunity of the circuit under design. This chapter discusses some timing related problems that influence the logic design level as well as the physical design level.

The link with testing is that a critical timing in most cases results in intermittent faults. These faults may have a very low probability of occurrence in a circuit, so they cannot be economically tested. The only way out is correct-by-design. But what does 'correct-by-design' entail? Recently, many papers have been published on all aspects of timing. Also, recent books on testing focus on timing and fault simulation [Abramovici90] and on timing verification [Rao89]. In this chapter we give some proposals for designing robust timing systems, starting at, and remaining restricted to the logic design level. In our conclusions we plead for a macro approach in clock system design as in logic and test design. We are then back to the macro approach, the theme in this book.

The delays in the components are not the only factor that determines the signal timing in high-speed digital circuits. The *transmission delays* in the wiring become as significant as gate delays. Design automation software is available to extract information about transmission delays in the wiring or the layout automatically. The *wiring delays* can be modeled as extra intrinsic delay elements in the wiring between the gates. For reasons of simplicity, we assume that wiring delays are modeled as a part of the gate delays. For time-critical applications a more detailed timing model should be used.

The same holds for the influence of the supply voltage on transmission delays. The supply voltage has a significant effect on the transmission and gate delays. As in CMOS switching currents are much higher than quiescent currents, the power supply has a static effect on the delays in a circuit as well as a dynamic effect. Much depends on the quality of the power distribution network on the chip or board. We assume that these effects are known and quantified in such a way that they can be modeled as extra gate delay on the logic design level as well. The same holds for the spread of parameters in production processes.

To be able to achieve the above we use a *minimum-maximum gate level timing model.* Under all environmental conditions, the worst case and the best case timing parameters are measured or obtained from simulations. In the following sections these minimum and maximum parameters are used as 'timing parameters'. To explain timing phenomena on the gate and the register-transfer level, this approach is sufficient.

A signal model for logic level timing analysis

In the static state, data signals in digital circuits are assumed to be in one of two *logic levels*. The logic level HIGH (H) is in an interval between a voltage level $V_{(min)}(H)$ and the supply voltage V_{DD}, while the logic level LOW is between $V_{(max)}(L)$ and ground. During a signal transition between the H and L levels the logic signal is said to be *undefined.*

The *threshold interval* lies somewhere between $V_{(max)}(L)$ and $V_{(min)}(H)$. Below the threshold, a circuit interprets its input signal definitely as L and above the threshold as H, under all permitted environmental conditions as required by the user or specified in a data book.

These definitions lead to the signal model given in Figure 8.1. A logic signal is said to be *defined* if it is in one of the logic levels L or H. In between, the signal has no logic value. It is *undefined.* In principle, the logic state of every component that is driven by an undefined signal is assumed to be undefined as well.

From Figure 8.1 it also follows that the *transition time* (rise or fall time) of the signal determines largely the duration of the interval in which the signal is undefined. This transition time must be obtained from measurements or simulations on the physical design level. For our purposes, we assume that the minimum and maximum transition time for every interconnection on a printed circuit board or in an IC is known. A more detailed timing model can always be applied.

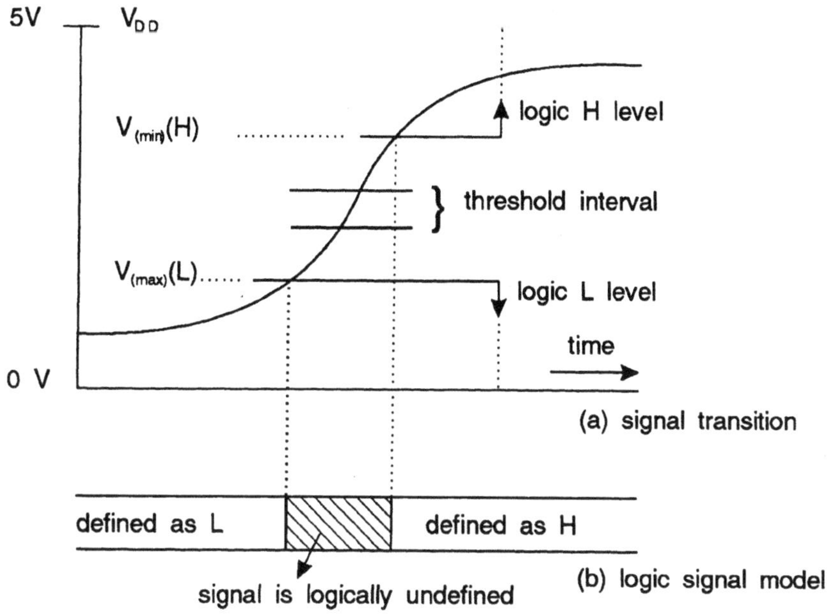

Figure 8.1. Defined/undefined logic signal model.

An important question, rarely referred to in the literature, is how and where to define the logic levels $V_{(min)}(H)$ and $V_{(max)}(L)$. Data books usually have a *reference level* V_{ref} for timing references. This is an easy way out. It leaves all interpretation problems to the designer.

Another approach is to refer to the levels $V_{in(min)}(H)$ and $V_{in(max)}(L)$, the minimum and maximum level at which a signal is guaranteed to be recognized as H and L at an input. Many papers, books and data books use these levels as reference levels. However, this approach does not take into account the DC and AC noise margins. In a digital circuit, these *noise margins* strongly determine the robustness and reliability of data transfers. If a logic signal is assumed to be present up to a time t, this is also implicitly assumed to be true for its AC and DC noise margins. The AC noise margin is the distance between the logic levels and the threshold interval. The DC noise margin is the distance in between $V_{in(min)}(H)$ and $V_{out(min)}(H)$ or $V_{in(max)}(L)$ and $V_{out(max)}(L)$. These margins are essential for the noise immunity of a circuit. Normally these margins are about 0.4 Volt. So:

$$V_{(min)}(H) = V_{out(min)}(H), \tag{8.1}$$

$$V_{(max)}(L) = V_{out(max)}(L). \tag{8.2}$$

The basis for the timing reference is not the input level, but the output level. Note that the output levels are closer to V_{DD} and ground, so the undefined signal intervals are longer, depending on the actual rise and fall time.

A logic level model for timing analysis

The gate model for timing analysis is based on the physical reaction of a gate output to a signal transition at any of its inputs. The gate output shows some delay before reacting to an input transition. Before $t_{P(min)}$(gate) the output does not show any response at all. This time is called the *minimum propagation delay time* of the gate. Up to $t_{P(min)}$(gate) the present output signal is available. For the parameter $t_{P(min)}$(gate) the best case conditions must be taken into account when performing simulations or measurements. After a time $t_{P(max)}$(gate) the output is again stable and in one of its logic levels. Now the worst case conditions must be taken into account to find the *maximum propagation delay time* of the gate.

A well-known property of gates is that the propagation delay time decreases when more than one input switches in the same direction. A reduction of the propagation delay time up to 50% is possible! Many data books specify the propagation delays for one input transition at a time.

Reference levels for an output reaction are output signal transitions that exceed the minimum and maximum levels $V_{out(min)}(H)$ and $V_{out(max)}(L)$. Output transitions above or below these levels are logically equivalent to no transitions at all.

Figure 8.2 shows the coherence between the parameters $t_{P(min)}$(gate) and $t_{P(max)}$(gate) and defined and undefined signal intervals at the inputs and outputs of a gate. We see that the undefined signal interval at the output of a gate is always longer than the corresponding undefined input interval.

For multi-level combinational circuits these effects are cumulative. We use a simple model, in which minimum and maximum propagation delays are simply added. An n-level circuit of gates with the same propagation delay parameters has n-fold values for the parameters. In practice, this problem is more complicated. The L-to-H transition time is not always the same as the H-to-L transition time.

A concatenation of two inverters, for example, has a shorter propagation delay time than twice $t_{P(max)}$(inverter). A low-level simulation can produce more realistic timing parameters.

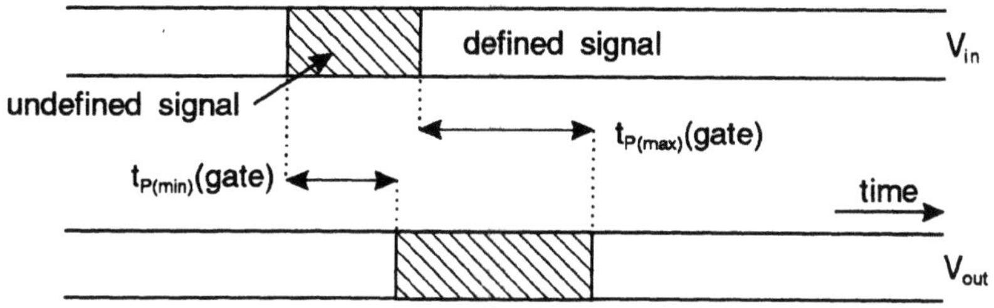

Figure 8.2. Timing model for logic gates.

False paths give also problems. A false path is a path through a combinational circuit that cannot be activated because of the signal values required for activation exclude each other. This phenomenon is well known and software for analysing false paths is available.

The minimum/maximum delay model is a *static delay model*. For many applications a more dynamically oriented delay model should be used. Take, for example, an AND gate with a low-level L (positive logic) at one of its inputs. What does the output of this gate do during signal transitions at the other inputs? Of course, you will assume that the output remains L. With the present-day fast technology this is not always true. Output transitions that exceed the levels $V_{out(max)}$(L) and $V_{out(min)}$(H) are frequently be found in simulations. Sometimes such transitions also exceed the threshold voltage of the driven gate. In the first case, the noise immunity of the circuit may become problematic during switching conditions. In the latter case, malfunctioning of the circuit may be the result.

It is the responsibility of the logic designer to point out those places in the design where a more dynamically oriented timing model is required. For example, all types of carry skip adders use a dynamic timing model in the skip path of the carry. During the silicon mapping, it must be guaranteed that the required dynamic properties are realized in the final circuit. Sometimes a redesign of a gate may be necessary!

8.2 Timing Models of Latches and Flip-Flops

Flip-flops are based on the master-and-slave principle. Usually a flip-flop is built with two gated latches. The timing parameters of the flip-flops depend on the timing parameters of the gated latches. Figure 8.3 shows the timing of a gated D latch.

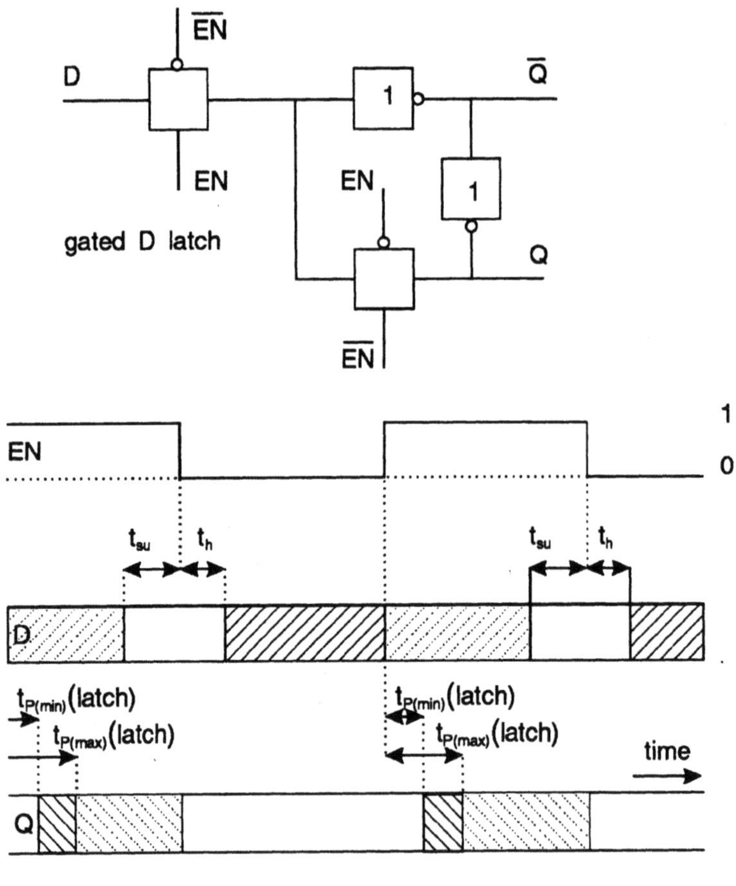

Figure 8.3. Timing of a gated D latch.

For EN = 0 the gated D latch is disabled. In this interval $Q_{new} = Q_{old}$. While EN = 1 the latch is enabled. Then Q_{ncw} = D, and the latch is transparent for the input data.

In order to give Q a predefined value during EN = 0, the D signal must be present during the 1→0 transition of the EN signal. The exact interval in which D must be present is defined by two timing parameters t_{su}(latch) = t_{su}(D-to-EN) and t_h(latch) = t_h(D-to-EN), the setup time and the hold time of D to EN.

At the 0→1 transition of EN the output of the gated latch starts following the D signal. The first reaction will be within the minimum and maximum propagation time of the gated D latch, $t_{P(min)}$(latch) = $t_{P(min)}$(EN-to-Q) and $t_{P(max)}$(latch) = $t_{P(max)}$(EN-to-Q). When the D signal only changes during EN = 0, this transition is the only output reaction of the latch. Else, the gated latch starts following the D input signal during EN = 1.

A preparatory setup of D during EN = 0 is the preferred way to set a new state in the latch, because of a lower dissipation in the latch. Then the latch output switches only once, in a *predefined interval.* If a somewhat higher dissipation is no problem, the interval in which the latch can be set into a new internal state and output value is longer. *So there is a possibility to exchange timing margins against power dissipation!*

Figure 8.4 shows the IEC symbol for a gated latch. This symbol corresponds to the circuit diagram in Figure 8.3.

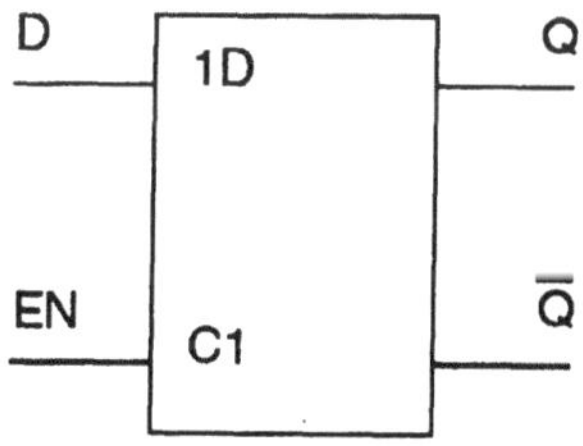

Figure 8.4. Symbol for a gated latch.

Flip-flop timing is based on the timing of the latches with which the flip-flop has been built. Figure 8.5 shows a flip-flop with its timing diagram. The input latch Y is enabled during $\alpha = 0$ and the output latch Z during $\alpha = 1$. As a result, the interval of the output transition, if any, is coupled with the 0-to-1 transition of the clock signal α. Such a flip-flop is called a *positive edge-triggered flip-flop*.

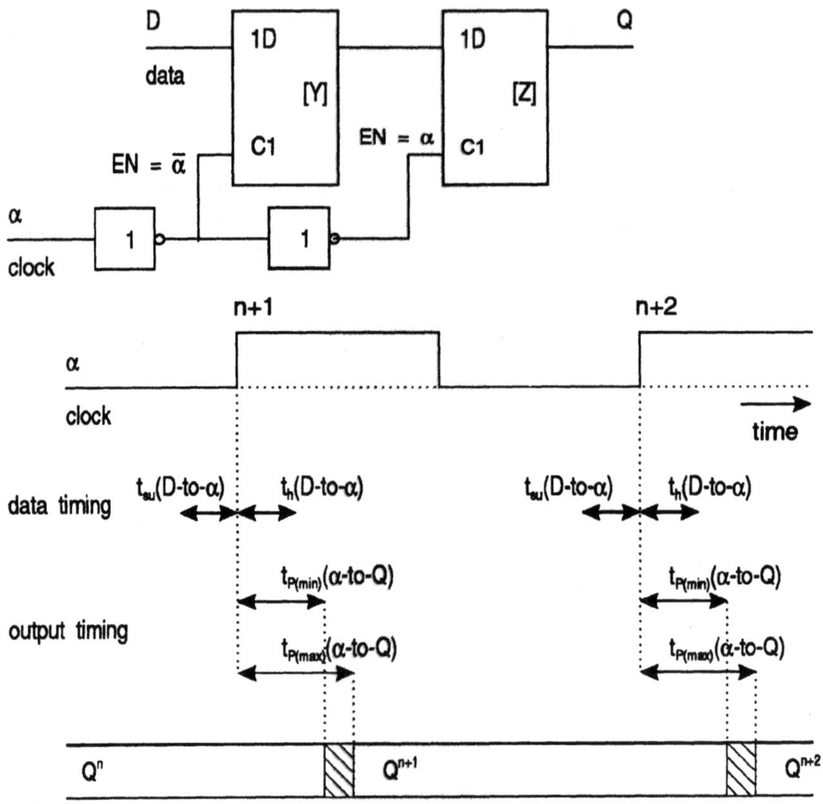

Figure 8.5. Timing specification of a positive edge-triggered D flip-flop.

An edge-triggered flip-flop has a small setup and hold interval for its data signals. Outside this interval, the data signal(s) have no influence on the internal state of the flip-flop. This means a high noise immunity of the flip-flop for noise on the data lines! Older types of flip-flops sometimes have a *pulse-triggered timing*. Then the data must be present around the whole H interval of the clock (positive pulse-triggered flip-flops) or around the L interval (negative pulse-triggered flip-flops). As a result the noise immunity is bad [Thijssen89].

Asynchronous inputs on flip-flops, such as an asynchronous set or reset, are a real disaster for design reliability. Those inputs must have a predefined logic level throughout the entire clock period. This means that their noise immunity is low. Measurements on printed circuit boards have shown that the error probability of an asynchronous input on a circuit may be about 1 at 10^8 clock pulse intervals. Much too high for digital circuits!

In modern IC designs only *preparatory inputs* are allowed, with a strong preference for an edge-triggered timing. In practice, it may be difficult to see if a signal on a flip-flop has an edge-triggered or a pulse-triggered timing, as the following example shows.

Example

Figure 8.6 shows a positive edge-triggered D flip-flop with an input selector. There are two data signals D_1 and D_2. Two select signals a and b are available to set the selector for data on D_1 (ab = 10), for data on D_2 (ab = 01) or for the inhibit mode of the flip-flop (ab = 00). From time to time this construction is found in CMOS integrated circuits.

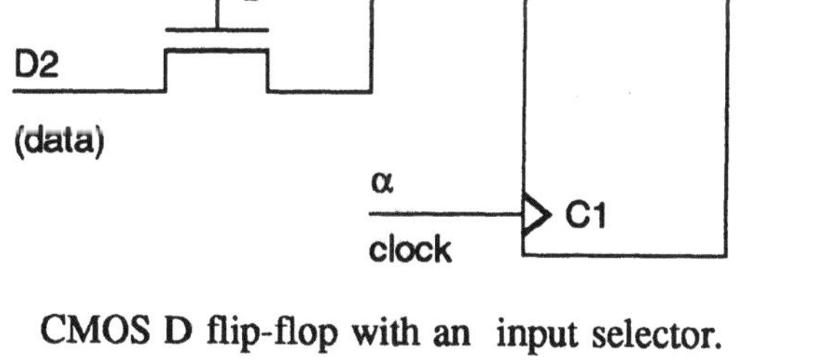

Figure 8.6. CMOS D flip-flop with an input selector.

The data signals have an edge-triggered timing with respect to the clock α. The select signals have an edge-triggered timing for ab = 10 or ab = 01. For ab = 00 (inhibit mode) the select signals a and b must always be kept 0 as long as the inhibit mode is active. As a result, the noise immunity is low during the inhibit mode.

Another objection to this construction is that for ab = 00 the flip-flop does not have a static behavior, but the flip-flop behaves as a dynamic flip-flop. As long as the flip-flop stores the electric charge at point p, the flip-flop state is maintained. But not longer!

A way out of the timing problem discussed above is to introduce a feedback path from the output Q of the flip-flop into the input selector. The inhibit function is realized by selecting the flip-flop output. For every pulse-triggered or asynchronous construction, a synchronous and/or edge-triggered realization can always be obtained. Sometimes such constructions require more transistors than asynchronous or pulse-triggered implementations. The price that must be paid for a reliable and noise-immune circuit!

8.3 Timing of data transfers

Figure 8.7 shows the timing of a data transfer between two positive edge-triggered D flip-flops. The parameters t_{su}(FF), t_h(FF), $t_{P(min)}$(FF) and $t_{P(max)}$(FF) are specified. We also assume that the transmission delay in the wiring has been included in the flip-flop parameters. Input data for flip-flop FF2 must be available in its setup and hold interval. The data is sent by flip-flop FF1, and is available outside the undefined intervals of its output. For correct data transfer, there must be no overlap between the setup-hold interval of FF2 and the undefined output interval of FF1. See Figure 8.7.

Any timing reference to a clock signal must take into account the clock's undefined intervals, in which the clock transitions take place. Clock transitions are always single transitions in the 'undefined' transition intervals. This uncertainty contributes to *clock skew*. Clock skew is the cumulative uncertainty at the moment of the clock transitions. Many phenomena contribute to clock skew. Examples are the following.

- Differences in clock generator outputs.
- (Capacitive) loads on each clock generator output.
- Restoring effects of clock buffers on transition times.
- Differences in logic levels in logic families, etc.

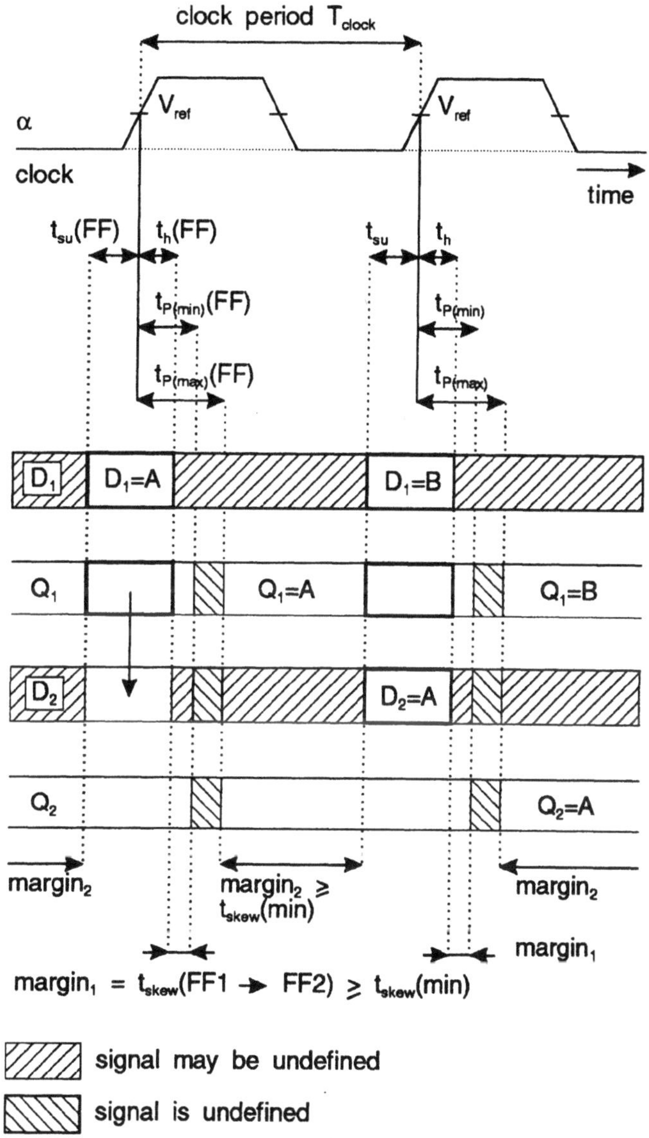

Figure 8.7. Direct data transfer between edge-triggered flip-flops.

A difference in the intrinsic propagation delay of clock signals in a circuit or system also contributes to clock skew. In this case we think of the following.

* Propagation delays in clock buffers.
* Propagation delays of clock signals in clock trees in the layout.
* Combinational logic in the clock path for switching clocks on and off.

Last but not least, the above phenomena depend on environmental parameters such as temperature, supply voltage and supply voltage drops during switching, tolerances in IC processes and IC aging processes. These phenomena have been studied in many papers [Bakoglu90, Chao92, Dagenais92, Nelis89]. Most papers concentrate on characterizing one of the phenomena or on keeping the influence of clock skew as low as possible. Evidently, there is always a certain minimum amount of clock skew. For every circuit an application range has to be defined, from which the *worst case margins* for clock skew have to be extracted. These activities result in a parameter $t_{skew(min)}$, denoting the *minimum margin that must be available* to compensate for clock skew. This margin must always be kept to meet the total amount of uncertainty in the clock reference.

We state explicitly that research on margins for clock skew by the authors has led to the conclusion that much higher requirements on margins for clock skew must be imposed on designs of digital ICs than most designers are taking into account. See the discussion on noise levels above and [Karlsson88].

Figure 8.7 shows the available margins for clock skew for direct data transfer from one edge-triggered D flip-flop to another edge-triggered D flip-flop. From this figure we see that there are two margins available, $margin_1$ and $margin_2$. The first margin, $margin_1$,

$$margin_1 = t_{P(min)}(FF1) - t_h(FF2) \qquad (8.3)$$

expresses that Q_1 is not allowed to change before the hold time t_h of input D_2 has passed. $Margin_2$,

$$margin_2 = T_{clock} - t_{P(max)}(FF1) - t_{su}(FF2) \qquad (8.4)$$

expresses that the next clock pulse must not be given too early. Both margins must be greater than specified by $t_{skew(min)}$. If not, reliable data transfer cannot be guaranteed.

The first margin, $margin_1$, only depends on flip-flop parameters. As such, the parameter $t_{skew}(FF)$,

$$t_{skew}(FF) = t_{P(min)}(FF) - t_h(FF) \quad (8.5)$$

is an important measure for the quality of a flip-flop design. It says how much margin is available for intercepting clock skew during data transfer to the same type of flip-flops. See Section 8.4 for additional information.

$Margin_2$ depends not only on flip-flop parameters, but also on the clock period T_{clock}. This margin can be made as wide as necessary by increasing the clock period. Only in high-frequency circuits may this margin become critical. With combinational logic between flip-flops the technique of retiming [Leiserson91] helps to solve the problem.

When the flip-flops shown in Figure 8.7 are in ICs of a different technology and flip-flop FF1 is faster than flip-flop FF2, the available margin

$$t_{skew}(FF1 \rightarrow FF2) = t_{P(min)}(FF1) - t_h(FF2) \quad (8.6)$$

is very critical. *The resulting margin for clock skew is smaller than it is with data transfers between flip-flops of the fast type only.*

Conclusions for direct data transfer are the following.

* For fast logic, the flip-flop dependent margin for clock skew $t_{skew}(FF)$ is very small and tends to be critical.
* The mixed logic situation, with different types of flip-flops in a design, is more critical than flip-flops in fast logic only. If possible, do not use different logic families in a design.
* If possible, a low-frequency technology is to be preferred because the greater margins.

Indirect data transfers

When flip-flop FF1 drives flip-flop FF2 by combinational logic, the margins for clock skew are:

$$margin_1 = t_{P(min)}(FF1) + t_{P(min)}(logic) - t_h(FF2) \quad (8.7)$$

$$margin_2 = T_{clock} - t_{P(max)}(FF1) - t_{P(max)}(logic) - t_{su}(FF2) \quad (8.8)$$

in which $t_{P(min)}$(logic) and $t_{P(max)}$(logic) are the timing parameters of the combinational logic between the flip-flops. Again, one of the margins is *frequency independent* (8.7), while the other margin is *frequency dependent* (8.8).

Additional logic increases the frequency independent margin for clock skew. As such, extra logic can be used to compensate for critical flip-flops or for critical data transfer:

$$\text{margin}_1^* = \text{margin}_1(\text{direct data transfer}) + t_{P(min)}(\text{combinational logic}) \quad (8.9)$$

$$\text{margin}_2^* = \text{margin}_2(\text{direct data transfer}) - t_{P(max)}(\text{logic}) \quad (8.10)$$

Another conclusion is that:

* wide margins for clock skew and a high performance of a circuit, e.g. a high clock frequency, can never be optimized together. In time-critical applications, the priority order of margins for clock skew (= reliability) and a high clock frequency (= performance) of the circuit under design must be stated in advance. The same conflict is found in flip-flop designs. See the next section.

8.4. Flip-flop designs

The previous sections showed why margins for clock skew are important. The most critical margin was the margin described by the flip-flop parameter t_{skew}(FF). In this section we demonstrate how the internal clocking structure of a flip-flop influences the flip-flop parameters and, as such, the performance and the margins for clock skew.

Performance

Figure 8.8 gives the internal structure of the conventional master-and-slave flip-flop. This is a single-phase clocked flip-flop. For this flip-flop the timing parameters are as follows.

$$t_{su}(FF) = t_{su}(Y\ latch) \tag{8.11.a}$$

$$t_h(FF) = t_h(Y\ latch) \tag{8.11.b}$$

$$t_{P(min)}(FF) = t_{P(min)}(INV) + t_{P(min)}(Z\ latch) \tag{8.11.c}$$

$$t_{P(max)}(FF) = t_{P(max)}(INV) + t_{P(max)}(Z\ latch) \tag{8.11.d}$$

It is easy to find the margins for clock skew from these parameters.

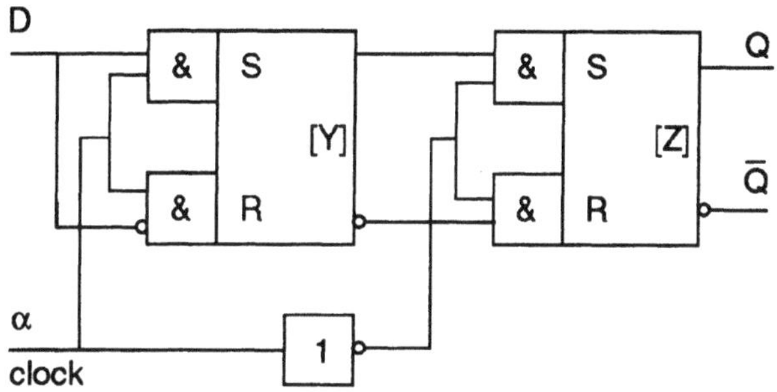

Figure 8.8. Traditional master-and-slave flip-flop.

The Figures 8.9 and 8.10 give two alternatives for the internal flip-flop structure. The flip-flop in Figure 8.9 has asymmetrical latches [Weste93], while the flip-flop in Figure 8.10 has a feedback clock on the input latch.

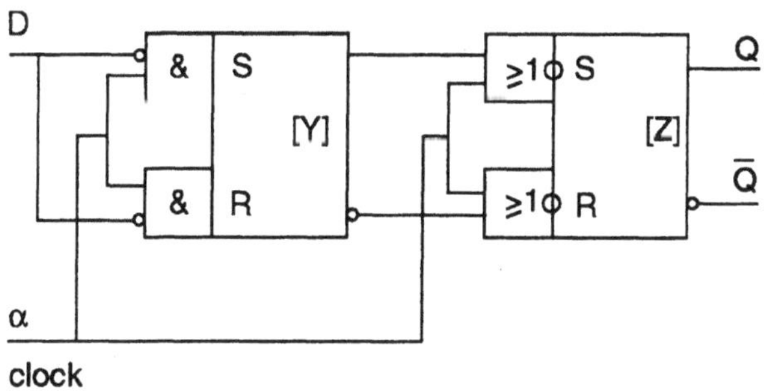

Figure 8.9. Flip-flop design.

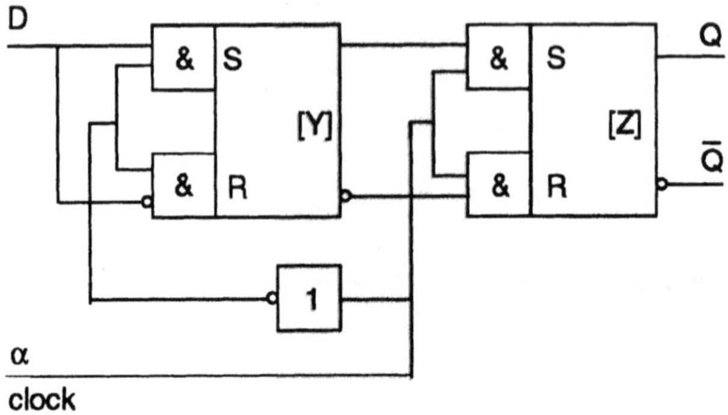

Figure 8.10. Flip-flop design.

A comparison with the flip-flop structure of Figure 8.8 shows that the flip-flop in Figure 8.9 fits better in high-performance designs. The *minimal clock period* is:

$$T_{clock(min)} = t_{P(max)}(\text{Z latch/FF1}) + t_{P(max)}(\text{logic}) + t_{su}(\text{Y latch/FF2}) + t_{skew(min)} \quad (8.12)$$

We see a gain of $t_{P(max)}(\text{INV})$, compared to $T_{clock(min)}$ for the design in Figure 8.8. The design in Figure 8.10 is even better. Now:

$$T_{clock(min)} = t_{P(max)}(\text{Z latch/FF1}) + t_{P(max)}(\text{logic}) + t_{su}(\text{Y latch/FF2}) - t_{P(min)}(\text{INV}) + t_{skew(min)} \quad (8.13)$$

Theoretically, the gain for the minimum clock period is two times t_{PD}(INV). From the *performance point of view*, the ranking of the internal flip-flop structures is: Figures 8.10 - 8.9 - 8.8.

Example

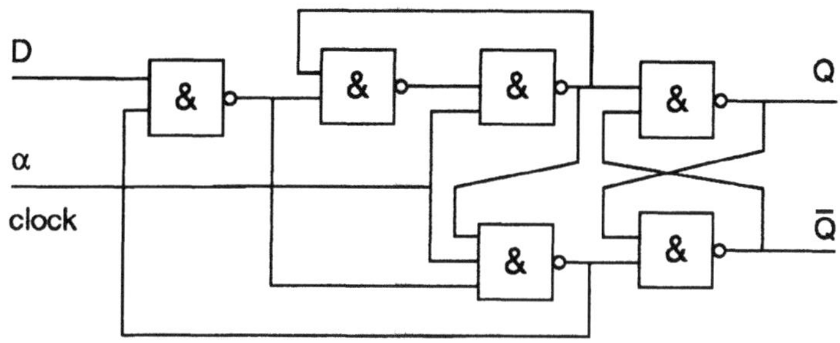

Figure 8.11. Structure of a TTL D flip-flop.

The D flip-flop as shown in Figure 8.11, the well-known TTL D flip-flop in NAND gates, has an internal structure comparable to the design model in Figure 8.10. The clock is directly fed to the output latch and has a feedback path to the input latch. A typical high-performance design.

Margins for clock skew

We find a completely opposite ranking order when observing the flip-flop dependent margins for clock skew. For the parameter t_{skew}(FF) the structures of the Figures 8.8 - 8.10 give the following results:

Figure 8.8: $t_{skew}(FF) = t_{P(min)}(INV) + t_{P(min)}(Z\ latch) - t_h(Y\ latch)$ (8.14)

Figure 8.9: $t_{skew}(FF) = t_{P(min)}(Z\ latch) - t_h(Y\ latch)$ (8.15)

Figure 8.10: $t_{skew}(FF) = t_{P(min)}(Z\ latch) - t_{P(max)}(INV) - t_h(Y\ latch)$ (8.16)

The internal structure shown in Figure 8.8 gives the greatest margin for clock skew, that of Figure 8.10 the shortest. Thus the ranking order for the margin t_{skew}(FF) is: Figures 8.8 - 8.9 - 8.10.

The flip-flop design conclusions are as follows.

* The design objective of maximal margins for clock skew is in conflict with the design objective of fhigh-performance. The priority order is obvious: reliability first. This means choosing for the widest margins for clock skew.

Clock skew margins in flip-flops have recently become a design problem because of the higher switching frequencies of flip-flops in fast CMOS technologies.

Two-phase clock systems

There are some other means that can help to compensate for clock skew. One of them is to place additional combinational logic between flip-flops. The margin t_{skew}(FF) enlarges by $t_{P(min)}$(logic). More freedom for clock skew can de obtained by using a *two-phase clock system*. Figure 8.12 gives an introduction.

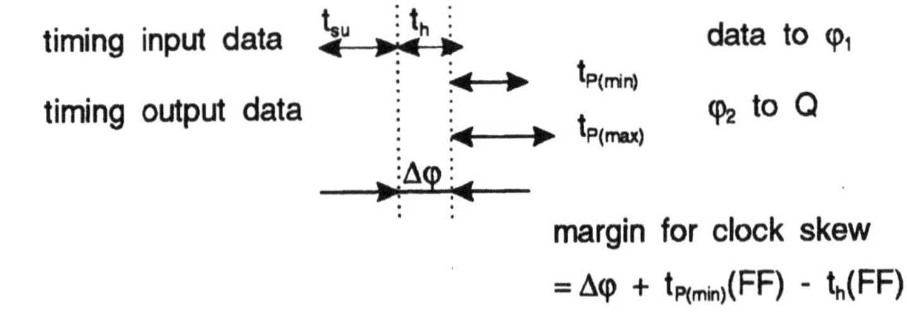

Figure 8.12. Two-phase clock system.

In a two-phase clock system there are two non-overlapping clock phases φ_1 and φ_2. Any margin for clock skew can be obtained by introducing a phase shift Δ between φ_1 and φ_2:

$$\Delta(\varphi_1\downarrow \rightarrow \varphi_2\uparrow) \quad (8.17)$$

This phase shift is normally realized in the clock generator. For high-speed circuits the preferred solution!

Another solution to the clock skew problem is found in Boundary Scan [Maunder90]. Data on the external scan line must be set at the trailing edge of the clock. This data is read at the leading edge of the clock in the next circuit. Thus almost half of the clock period is available as margin for clock skew. Another effect is that both margins for clock skew have become frequency dependent, via the pulse/pause ratio of the clock. This timing system is known as *data lock-out timing*. For general use, the overhead is too high, but for special applications, such as in Boundary Scan, it is worth considering.

8.4 Clock drivers

Normally, in a clock distribution network *clock drivers* are used. Their function is to increase the number of components that can be driven by a clock signal. Another application is at the inputs of ICs, to restrict the fan in of external input signals. Many older SSI/MSI circuits and the recent large ICs have on-chip clock drivers at the clock input. Figure 8.13 shows a design with circuits with a buffered clock input and circuits with no clock drivers. For simplicity, we assume all flip-flops are of the same type and have identical parameters.

Clock drivers delay the clock. They have a negative influence on the frequency independent margins for clock skew. In data transfer, the frequency independent margins are:

FF1 from/to FF2:

$$\text{margin} = t_{P(min)}(FF) - t_h(FF) \quad (8.18)$$

$$= t_{skew}(FF \rightarrow FF)$$

FF1 to REG:

$$\text{margin} = t_{P(min)}(FF) - t_h(FF) - t_{P(max)}(\text{clock driver}) \qquad (8.19)$$

$$= t_{skew}(FF \rightarrow FF) - \mathbf{t_{P(max)}(clock\ driver)}$$

REG to FF2:

$$\text{margin} = t_{P(min)}(FF) - t_h(FF) + t_{P(min)}(\text{clock driver}) \qquad (8.20)$$

$$= t_{skew}(FF \rightarrow FF) + \mathbf{t_{P(min)}(clock\ driver)}$$

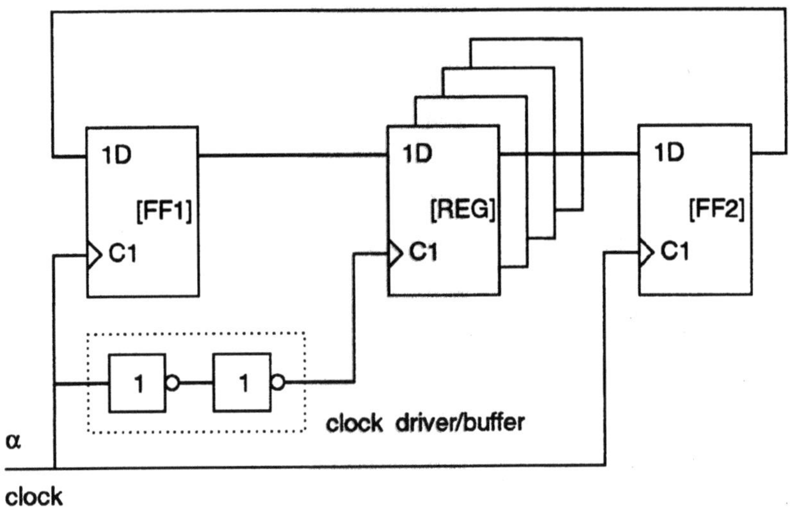

Figure 8.13. Data transfer with buffered registers.

The conclusions for clock buffers/drivers are as follows.

* Clock buffers/drivers with different propagation delays claim a part of the frequency independent clock skew margin. As this margin is generally very small, extra precautions must be made. A recent way of solving this problem in large ICs is to use phase-locked loop clock drivers.
* The propagation delay of clock buffers/drivers can be compensated for by using the propagation delay of extra logic in front of the data inputs of

the buffered flip-flops. The spread in propagation delay of this logic must also be taken into account as clock skew.

In ultra-large IC designs, the transmission time in the *clock trees* can no longer be neglected. In principle, all measures discussed above can be applied. To restrict clock skew, clock trees are balanced. This can be done by introducing an extra load in a branch, if the load is lower than in other branches of the clock trees. As such, this is a static way to compensate for clock skew. Clock skew has also a dynamic component. During switching conditions, the load a flip-flop imposes on a clock line is dependent on the data: does the flip-flop change its internal state or not? Also, more and more circuits have clock switches that can switch on/off clocks to reduce the dissipation. This has also a dynamic effect on the load of some branches in a clock tree.

Dynamic effects are very difficult to quantify in simulations. Sometimes designers choose a point-to-point approach. They verify the timing by simulating every data transmission path in a circuit. When found to be necessary, extra precautions are taken to compensate for clock skew. A better approach is to define *clock macros* during the design phase of a circuit. A clock macro is a part of a circuit that is at the same branch of a clock tree. Within the macro, flip-flops are of the same type, with a predefined margin for clock skew. There are no clock skew problems within a macro!

Clock macros are chosen such that there are as few data lines between different clock macros as possible. Thus the problems concentrate on a relatively small number of interconnections. Clock skew can then easily be compensated. It must explicitly be stated that clock macros do not necessarily have the same boundaries as logic macros or test macros.

In this chapter we have studied some 'high-level' aspects of clocking and clock skew. We started with a discussion on the definition of the reference levels for the timing. Then we went into the details of the flip-flop design to create as much margin for clock skew as possible. After which we presented some methods to increase the margins for clock skew with extra logic, or with a two-phase clock. Finally, we ended with advicing for a divide-and-conquer strategy. In doing this we restricted ourselves to the *register-transfer level*. The concepts must be translated to the physical desing level. There the exact values of the parameters are known, and there it can also be verified if a certain approach suffices or not. It is our opinion that all that can be done at the logic level to improve reliability and noise immunity should be done at that level.

List of Symbols and Abbreviations

Symbols

$\mathbf{N}$	set of natural numbers
$\wp(\mathbf{N})$	power set of the natural numbers
$\lceil x \rceil$	$\min\{ n \in \mathbf{N} \mid n \geq x \geq 0\}$
Rd(0)	read instruction of a memory cell expecting logical value 0
Rd(1)	read instruction of a memory cell expecting logical value 1
Wr(0)	write instruction for a logical value 0 to the addressed memory cell
Wr(1)	write instruction for a logical value 1 to the addressed memory cell
N	number of memory addresses
BN	set of bit error numbers
SB: BN $\rightarrow$ {0,1,2}	step function defining the signature bit
ST_scantest	control signal for Built-In Self-Test
ST_runtest	control signal for Built-In Self-Test
B	set of logic blocks
m	number of logic blocks
SR	set of scan registers
sr	scan register, sr $\in$ **SR**
n	number of scan registers
T	set of tests to be applied to a device
t	set of leaf macro test patterns, t $\in$ **T**
s	number of tests in **T**
$N(t_i)$	number of test patterns in t_i
$RA(t_i)$	set of pattern-applying scan registers for test t_i
$RO(t_i)$	set of pattern-observing scan registers for test t_i
SCC	set of scan chain configurations
scc	scan chain configuration, scc $\in$ **SCC**
k(scc)	number of scan chains in scc
SO(n,k)	number of possible scan chain configurations of n scan registers with k scan chains
C(scc)	cost of scan chain configuration scc.
$\check{C}$(sc, t)	cost of scan chain sc for test t
$\hat{C}$(scc,t)	cost of scan chain configuration scc for test t
$sr_i \rightarrow sr_j$	ordering relation of two scan registers in the same scan chain

sc	scan chain, sc = (sr_p, ..., sr_q) with $sr_p \rightarrow sr_{p+1} \rightarrow ... \rightarrow sr_q$
d^{in}(sr,sc)	input distance of scan register sr in scan chain sc
d^{out}(sr,sc)	output distance of scan register sr in scan chain sc
d'^{in}(sc,t)	input distance of scan chain sc and test t
d'^{out}(sc,t)o	utput distance of scan chain sc and test t
l(sr)	length of scan register sr
l_i	shorthand notation for length of scan register sr_i
N^{in}(sr)	number of times scan register sr is used as pattern-applying register
N^{out}(sr)	number of times scan register sr is used as pattern-observing register
Apply(sr,t)	function indicating whether sr ∈ RA(t)
Observe(sr,t)	function indicating whether sr ∈ RO(t)
F^{in}(sc)	set of farthest input scan registers
F^{out}(sc)	set of farthest output scan registers
N^{fin}(sc,sr)	number of times scan register sr in sc is used as farthest input register
N^{fout}(sc,sr)	number of times scan register sr in sc is used as farthest output register
MaxApply(sr,t)	function indicating whether sr is farthest input scan register for test t
MaxObserve(sr,t)	function indicating whether sr is farthest output scan register for test t
W(sr)	weight of scan register sr
I	finite, non empty set of inputs for a finite state machine
I^*	set of all finite input sequences that can be constructed by concatenating the elements of I
O	finite, non empty set of outputs of a finite state machine
S	finite, non empty set of internal states of a finite state machine
δ IxS $\rightarrow$ S	state transition function of a Moore Machine
δ. I^*xS $\rightarrow$ S	extended state transition function
λ S $\rightarrow$ O	output function of a Moore Machine
λ I^*xS $\rightarrow$O	extended output function
$s_i \equiv s_j$	equivalence relationship between two states s_i, s_j ∈ S
s_R	reset state
s_d	unused state
TC	set of test control signals
tc	test control signal, tc ∈ TC
TB	set of Test Control Blocks
tc-scope(tc)	scope of test control signal tc
tcb-scope(tcb)	scope of Test Control Block tcb

GTP	set of generated test plans
gtp	generated test plan, gtp $\in$ **GTP**
R	set of available resources
R^{in}(gtp)	input resources for generated test plan gtp
R^{out}(gtp)	output resources for generated test plan gtp
R^{con}(gtp)	control resources for generated test plan gtp
R(gtp)	resources for test plan gtp
Time(gtp,r)	set of cycles at which r is used as a resource for gtp
Value(gtp,r,n)	value assigned to r at cycle during execution of gtp
$r_1 \varnothing r_2$	time-disjoint property of resources r_1 and r_2
$r_1 \sqsubset r_2$	cover property of resources r_1 and r_2
$gtp_1 \sim_{tp} gtp_2$	test plan compatibility of generated test plans gtp_1 and gtp_2
$gtp_1 \sim_{tps} gtp_2$	test plan step compatibility of generated test plans gtp_1 and gtp_2
$gtp_1 \sim_{tss} gtp_2$	test specification step compatibility of generated test plans gtp_1 and gtp_2

Abbreviations

ATE	Automatic Test Equipment
BIST	Built-In Self-Test
DDD	Defect Density Distribution
DfT	Design-for-Testability
DSD	Defect Size Distribution
DRAM	Dynamic Random Access Memory
DSP	Digital Signal Processing
EDIF	Electronic Design Interchange Format
FS	Fault Simulation
IC	Integrated Circuit
IFA	Inductive Fault Analysis
LFSR	Linear Feedback Shift Register
LMSD	Leaf Macro State Diagram
MISR	Multiple Input linear feedback Shift Register
MTM	Macro Test Mode
OSF	Open Software Foundation
PCB	Printed Circuit Board
PCM	Process Control Module
PLA	Programmable Logic Array
ROM	Read Only Memory
SCRP	Scan Chain Routing Problem
SCRP	Decision Variant of Scan Chain Routing Problem

SEM	Scanning Electron Microscope
SRAM	Static Random Access Memory
TAP	Test Access Port
TCB	Test Control Block
TCB FSM	Test Control Block Finite State Machine
TCK	Test ClocK
TDI	Test Data In
TDO	Test Data Out
TMS	Test Mode Select
TPG	Test Pattern Generation
TSF	Test Specification Format
VLSI	Very Large Scale Integration
VMX	VLSI MIX monitor

References

[Abadir83] M. Abadir, H. Reghbati, "Functional Testing of Semiconductor Random Access Memories", A.C.M. Computing Surveys, Vol. 15, no.3 Sept.1983 , pp. 174-198.

[Abadir85] M. S. Abadir and M.A. Breuer, "A knowledge-based system for designing testable VLSI chips", IEEE Design and Test of Computers, Aug. 1985, Vol. 5, No. 3, pp. 56-68.

[Abadir86] M.S. Abadir, M.A. Breuer, "Test Schedules for VLSI Circuits Having Built-In Test Hardware", IEEE Transactions on Computers, April 1986, pp. 361-367.

[Abadir89] M.S. Abadir, "TIGER: Testability Insertion Guidance Expert System", Proceedings IEEE International Conference on Computer-Aided Design, pp. 562-565.

[Abramovici90] M. Abramovici, M.A. Breuer, A.D. Friedman, "Digital Systems Testing and Testable Design", Computer Science Press, New York, 1990.

[Agrawal87] Vishwani D. Agrawal, Kwang-Ting Cheng, Daniel D. Johnson, Tonysheng Lin, " Complete Solution to the Partial Scan Problem", Proc. International Test Conference, 1987, pp. 44-51.

[Agrawal93a] Vishwani D. Agrawal, Charles R. Kime, Kewal K. Saluja, "A Tutorial on Built-In Self-Test, Part 1: Principles", IEEE Design & Test of Computers, March 1993, pp. 73-82.

[Agrawal93b] Vishwani D. Agrawal, Charles R. Kime, Kewal K. Saluja, "A Tutorial on Built-In Self-Test, Part 2: Applications", IEEE Design & Test of Computers, June 1993, pp. 69-77.

[Baker93] K. Baker, R. Morren, M. Wahl, B. Verhelst, "Using EDIF in IC Testing: Experience from the EVEREST Project", Proc. 3rd European Test Conference, Rotterdam 1993, pp. 434-443.

[Bakoglu90] H.B. bakoglu, "Circuits, Interconnections and Packaging for VLSI", Addison-Wesley, Reading, Massachusetts, 1990.

[Bardell87] P.H. Bardell, W.H. McAnney, and J. Savir, "Built-In Test for VLSI: Pseudorandom Techniques", John Wiley & Sons, New York, 1987.

[Bassett90] Robert W. Bassett et. al., "Low-cost Testing of High-Density Logic Components", IEEE Design and Test, April 1990, pp. 15-28.

[Beenker86] F.P.M. Beenker, K.J.E. van Eerdewijk, R.B.W. Gerritsen, F.N. Peacock and M. van der Star, "Macro testing, Unifying IC and Board Test", IEEE Design and Test of Computers Magazine, Dec. 1986, pp. 26-32.

[Beenker89] F. Beenker, R. Dekker, R. Stans, M. van der Star, "A Testability Strategy for Silicon Compilers", IEEE Int. Test Conf., Washington, August 1989, pp. 660-696.

[Beenker90] F.P.M. Beenker, R.W.C. Dekker, R.J. Stans, M. van der Star, "Implementing Macro Test in Silicon Compiler Designs", IEEE Design and Test of Computers Magazine, April 1990, pp. 41-51.

[Beenker92] Frans Beenker, "Macro-Test: A VLSI testable-design technique, Algorithmic and Knowledge based CAD for VLSI", edited by Gaynor Tayler and Gordon Russell, Peter Peregrinus Ltd, 1992.

[Beenker94] Frans Beenker, "Testability Concepts for Digital ICs", PhD Thesis University Twente, The Netherlands, April 1994.

[Bennetts84] R.G. Bennetts, "Design of Testable Logic Circuits", Addison-Wesley, 1984.

[Bennetts93] Ben Bennetts and Frans Beenker, "Partial Scan: what Problem does it Solve?", Proceedings IEEE European Test Conference, 1993, pp. 99-106.

[Bhavsar81] D.K. Bhavsar and R.W. Heckelman, "Self-Testing by Polynomial Division", Proceedings of the IEEE International Test Conference 1981, pp. 208-216.

[Biggs85] Norman L. Biggs, "Discrete Mathematics", Oxfort University Press, 1985.

[Bouwman92a] Frank Bouwman, Steven Oostdijk, Rudi Stans, Ben Bennetts, Frans Beenker, "Macro Testability: The Results of Production Device Applications", Proc. IEEE International Test Conference, 1992, Baltimore,pp. 232-241.

[Bouwman92b] Frank Bouwman, "Optimisation and Specification of the Test Control Block", Master Thesis, Delft University of Technology, department of electrical engineering, report 1_68340_28(1992)02.

[Bouwmeestr93] Hans Bouwmeester, Steven Oostdijk, Frank Bouwman, Rudi Stans, Loek Thijssen, Frans Beenker, "Minimising Test Time by Exploiting Parallelism in Macro Test", Proceedings IEEE International Test Conference, 1993, pp. 451-460.

[Breuer88] Melvin A. Breuer, Rajiv Gupta, Jung-Cheun Lien, "Concurrent Control of Multiple BIT Structures", Proceedings International Test Conference, 1988, pp. 431-442.

[Bruls91] E. Bruls, F. Camerik, H. Kretschman, J. Jess, "A Generic Method to Develop a Defect Monitoring System for IC Processes", Proceedings IEEE International Test Conference, 1991, pp. 218-227.

[Bruls93] E. Bruls, "Reliability Aspects of Defect Analysis", Proc. 3rd European Test Conference, Rotterdam 1993, pp. 17-26.

[Carter82] W.C. Carter, "Signature Testing with Guaranted Bounds for Fault Coverage", Proceedings of the IEEE International Test Conference 1982, pp. 75-82.

[Catthoor89] F. Catthoor, J. van Sas, L. Inze, and H. de Man, "A Testability Strategy for Multiprocessor Architecture", IEEE Design & Test of Computers, April 1989, pp. 18-34.

[Chao92] T.H. Chao, Y.C. Hsu, J.M Ho, K.D. Boesse, and A.B. Kahng, "Zero Skew Clock Routing with Minimum Wirelength", IEEE Transactions on Circuits and Systems, Vol 39, 1992, pp. 799-814.

[Claasen89] T. Claasen, F. Beenker, J. Jamieson, B. Bennetts, "New Directions in Electronics Test Philosophy, Strategy and Tools", Invited paper for the 1st European Test Conference, Paris, April 1989, pp. 5-13.

[Craig88] L. Craig, C.R. Kime, K.K. Saluja, "Test Scheduling and Control for VLSI Built-In Self-Test", IEEE Transactions on Computers, September 1988, pp. 1099-1109.

[Dagenais92] M.R. Dagenais, S. Gaiotti, and N.C. Rumin, "Transistor-level Estimation of Worst-Case Delays in MOS VLSI Circuits", IEEE Transactions on Computer-Aided Design, Vol. 11, 1992, pp. 384-395.

[DasGupta84] S. DasGupta, M.C. Graf, R.A. Rasmussen, R.G. Walther, T.W. Williams, "Chip Partitioning Aid: A Design Technique for Partitionability and Testability in VLSI", Proceedings Design Automation Conference, 1984, pp. 203-208.

[Dekker87] Rob Dekker, "Fault Modeling and Self-Test of Static Random Access Memories", TUD report 1-68340-28(1987)25, Department of Electrical Engineering, Delft University of Technology, The Netherlands, 1987

[Dekker88a] R. Dekker, F. Beenker and L. Thijssen, "Fault Modeling and Test Algorithm Development for Static Random Access Memories", Proc. IEEE Int. Test Conference, Washington, September 1988, pp. 343-352.

[Dekker88b] R. Dekker, F. Beenker and L. Thijssen, "A Realistic Self-test Machine for Word Oriented Static Random Access Memories", Proc. IEEE Int. Test Conference, Washington, September 1988, pp. 353-361.

[DenBeste86] William Den Beste, "Tools for Test Development", VLSI Systems Design, July 1986.

[Eichelbrg78] E. Eichelberger, T.W. Williams, "A Logic Design Structure for LSI Testability", Journal of Design Automation and Fault-Tolerant Computing, Vol, 2, No. 2, May 1978, pp. 165-178.

[Eichelbrg91] E.B. Eichelberger, E. Lindbloom, J.A. Waicukauski, T.W. Williams, "Structured Logic Testing", Prentice Hall, 1991.

[Ferguson88a] F. Joel Ferguson and John P. Shen, "Extraction and simulation of realistic CMOS faults using Inductive Fault Analysis", Proceedings IEEE International Test Conference, 1988, 475-484.

[Ferguson88b] F.J. Ferguson, John P. Shen, "A CMOS Fault Extractor for Inductive Fault Analysis", IEEE Transactions on Computer-Aided Design, vol 7, no. 11, November 1988, pp. 1181-1194.

[Friedman73] A.D. Friedman, "Easily Testable Iterative Systems", IEEE Transactions on Computers, Vol C-22, Dec. 1973, pp. 1061-1064.

[Fujiwara85] Hideo Fujiwara, "Logic Testing and Design for Testability", The MIT Press, Cambridge, Massachusetts, 1985, pp. 206-237.

[Gelsinger87] Patrick Gelsinger, "Design and Test of the 80386", IEEE Design & Test of Computers magazine, June 1987, pp. 42-50.

[Goor91] A.J. van de Goor, "Testing Semiconductor Memories: Theory and Practice", Wiley & Sons,Chichester, England, 1991.

[Goor93] Ad J. van de Goor, "Using March Tests to Test SRAMs", IEEE Design & Test of Computers, March 1993, pp. 8-14.

[Gracio89] J.A. Gracio, P.A. Bicudo, N.N. Rua, A.M. Oliveira, C.F.B. Almeida, J.P. Teixeira, "Test Preparation and Fault Analysis using a Bottom-up Methodology", Proceedings 1st European Test Conference 1989, pp. 168-174.

[Guterl84] Fred Guterl, "In Pursuit of the One-Month Chip", IEEE Spectrum, September 1984, pp. 28-49.

[Hall67] M. Hall, Jr. "Combinatorial Theory", John Wiley & Sons, Inc., 1967.

[Hapke89] Friedrich Hapke, "Automatic Test Program Generation for a Block Oriented VLSI Chip Design", Proc. European Test Conference, Paris, 1989, pp. 71-76.

[Hartmanis66] J. Hartmanis and R.E. Stearns, "Algebraic Structure Theory of Sequential Machines", Englewood Cliffs, Prentice-Hall, 1966.

[Hayes75] J. Hayes ,"Detection of Pattern Sensitive Faults in Random Access Memories", IEEE Transactions on Computers , Vol. C-24, no.2, pp. 150-157, Feb.1975.

[Hollmann90] Henk Hollmann, "Design of Test Sequences for VLSI Self-Testing Using LFSR", IEEE Transactions on Information Theory, vol 36, no. 2, March 1990, pp. 386-392.

[IEEE90] IEEE Standard 1149.1--1990, "IEEE Standard Test Access Port and Boundary-Scan Architecture", IEEE, New York 1990.

[Jone89] W.B. Jone, C.A. Papachristou, M. Pereira, "A Scheme For Overlaying Concurrent Testing of VLSI Circuits", Proceedings 26th Design Automation Conference, 1989, pp. 531-536.

[Karlsson88] I. Karlsson, "True Single Phase Clock Dynamic CMOS Circuit Technique", Int. Symposium on Circuits and Systems, ISCAS 1988, pp. 475-478.

[Kime82] Charles R. Kime, Kewal K. Saluja, "Test Scheduling in Testable VLSI Circuits", Proceedings International Symposium on Fault-Tolerant Computing, 1982, pp. 406-412.

[Knaizuk77] J. Knaizuk and C. Hartman, "An Optimal Algorithm for Testing Stuck-at Faults in Random Access Memories", IEEE Transactions on Computers, 1977, C-26, pp. 1141-1144.

[Konemann79] B. Konemann, J. Mucha, G. Zwiehoff, "Built-In Logic Block Observation Technique", Proceedings of the IEEE International Test Conference, 1979, pp. 37-41.

[Kuban84] John Kuban and John Salick, "Testability Features of the MC68020", Proceedings of the IEEE International Test Conference, 1984, pp. 821-826.

[Lee90] Sunggu Lee, Kang G. Shin, "Design for Test using Partial Parallel Scan", IEEE Transactions on Computer-Aided Design, February 1990, pp. 203-211.

[Leenstra90] Jens Leenstra, Lambert Spaanenburg, "Hierarchical Test Assembly for Macro Based VLSI Design", Proceedings International Test Conference, 1990, pp. 520-529.

[Leenstra93a] Jens Leenstra, "Hierarchical Test Development and Design-For-Testability for (A)synchronous Semi-Custom ASICs", PhD thesis Technical University Eindhoven, 1993.

[Leenstra93b] Jens Leenstra, Michael Koch and Thomas Schwederski, "On Scan Path Design for Stuck-Open and Delay Fault Detection", Proc. 3rd Europena Test Conference, Rotterdam 1993, pp. 201-210.

[Leiserson91] Ch.E. Leiserson, and J.B. Saxe, "Retiming Synchronous Circuitry", Algorithmica, 1991, pp. 5-25.

[Ligthart86] M.M. Ligthart, E.H.L. Aarts and F.P.M. Beenker, "Design for Testability of PLA's using Statistical Cooling", Proc. Design Automation Conference, Las Vegas 1986, pp. 339-345.

[Lunde89] R. Lunde and M. Faust, "Separating Timing, Data, and Format in a Tester Independent Waveform Representation", Proceedings of the European Test Conference, Paris, 1989, pp. 377-382.

[Ma88] Hi-Keung Tony Ma, Srinivas Devadas, A. Richard Newton, Alberto Sangiovanni-Vincentelli, "An Incomplete Scan Design Approach to Test Generation for Sequential Machines", Proc. International Test Conference, 1988, pp. 730-734.

[Maly85] W. Maly ,"Modeling of Lithography Related Yield Losses for CAD of VLSI Circuits", IEEE Transactions on CAD, Vol. CAD-4, no.3, pp. 166-177, July 1985.

[Maly86] W. Maly, "Fault Models for the NMOS Programmable Logic Array", IEEE Custom Integrated Circuits Conference, 1986, pp. 467-470.

[Maly88] W. Maly and P. Nigh, "Built-In Current Testing Feasibility Study", Proceedings ICCAD, pp. 340-343, Santa Clara, 1988.

[Maly89] W. Maly and S. Naik, "Process Monitoring Oriented Testing", Proc. International Test Conference, 1989, pp. 527-532.

[Maly90] Wojciech Maly, "Computer-Aided Design for VLSI Circuit Manufacturability", Proceedings of the IEEE, Vol. 78, No. 2, February 1990, pp. 356-392.

[Marinescu82] M. Marinescu ,"Simple and Efficient Algorithms for Functional RAM Testing", Proceedings of the IEEE International Test Conference 1982, pp. 236-239.

[Matterne87] L. Matterne, J. van Meerbergen, F. Beenker, V. Mehra, J. Theunissen, R. Segers, "A C-testable Booth Multiplier Designed in a Silicon Compilation Environment", Proc. IEEE Int. Conf. on Computer Design, 1987.

[Marinissen90] Erik Jan Marinissen, "Automated Test Control Block Generation and Minimisation", Master's Thesis Eindhoven University of Technology, Department of Mathematics and Computing Science, 1990.

[Marinissen91] Erik Jan Marinissen, Rob Dekker, "Minimisation of Test Control Blocks", Proceedings 2nd European Test Conference, Muenich, 1991.

[Maunder90] C.M. Maunder, and R.E. Tullos, "The Test Access Port and Boundary-Scan Architecture", IEEE Computer Society Press, Los Alamitos, 1990.

[Maxwell92] Peter C. Maxwell, Robert C. Aitken, Vic Johansen, Inshen Chiang, "The Effectiveness of I_{DDQ}, Functional and Scan Tests: How many Fault Coverages do we need?", Proceedings IEEE International Test Conference, 1992, pp. 168-177.

[McCluskey84] Edward J. McCluskey, "A Survey of Design for Testability Scan Techniques", VLSI Design, December 1984, pp. 38-61.

[Mealy55] G.H. Mealy, "A Method for Synthesising Sequential Circuits", Bell Syst. Techn. Journal, Vol. 34, September 1955, pp. 1045-1079.

[Meershoek90] Rob Meershoek, Bas Verhelst, R. McInerney, L. Thijssen, "Functional and I_{DDQ} Testing of Static RAM", Proceedings IEEE International Test Conference 1990, pp. 929-937.

[Mehtani90] R. Mehtani, K. Baker, C.M. Huizer, P.J. Hynes, J. van Beers, "Macro-Testability and the VSP", Proceedings International Test Conference, 1990, pp. 739-748.

[Mehtani92] R. Mehtani, M. de Jonghe, R. Morren and K. Baker, "Improving Total IC Design Quality using Application Mode Testing", Proceedings IEEE International Test Conference 1992, pp. 866-872.

[Moore64] E.F. Moore, 'Sequential Machines: Selected Papers, Reading", Addison-Wesley, 1964.

[Muehldorf82] E.I. Muehldorf, T.W. Williams, "Analysis of the Switching Behavior of Combinatorial Logic Networks", Proceedings International Test Conference 1982, pp. 379-390.

[Murray88] Brian T. Murray, John P. Hayes, "Hierarchical Test Generation Using Precomputed Tests for Modules", Proceedings IEEE International Test Conference, 1988, pp. 221-229.

[Naik93] Samir Naik, Frank Agricola, Wojciech Maly, "Failure Analysis of High-Density CMOS SRAMs Using Realistic Defect Modeling and I_{DDQ} Testing", IEEE Design & Test of Computers, June 1993, pp. 13-23.

[Nair78] R. Nair, S. Thatte and J. Abraham ,"Efficient Algorithms for Testing Semiconductor Random-Access Memories", IEEE Transactions on Computers, Vol. C-27, No. 6, June 1978, pp. 572-576.

[Nair79] R. Nair ,"Comments on An Optimal Algorithm for Testing Stuck-at Faults in Random Access Memories", IEEE Transactions on Computers, Vol. C-28, No. 3, March 1979, pp. 258-261.

[Narayanan92] S. Narayanan, C. Njinda, M. Breuer, "Optimal Sequencing of Scan Registers", Proc. International Test Conference, 1992, pp. 293-302.

[Nelis89] H. Nelis, P. Dewilde, and E. Deprettere, "Efficient Modeling of Interconnections in a VLSI Circuit", Int. Symposium on Circuits and Systems, 1989, pp. 872-876.

[Oostdijk91] S. Oostdijk, F. Beenker and L. Thijssen, "A Model for Test-time Reduction of Scan-testable Circuits", 2nd European Test Conference, Muenich, April 1991.

[Ostapko79] D.L. Ostapko and S.J. Hong, "Fault Analysis and Test Generation for PLAs", IEEE Transactions on Computers, vol. c-28, no. 9, September 1979, pp. 617-627.

[Papachristou85] C. Papachristou and N. Sahgal ,"An Improved Method for Detecting Functional Faults in Semiconductor Random Access Memories", IEEE Transactions on Computers, Vol. C-34, no.2, February 1985, pp. 110-116.

[Paull59] M.C. Paull and S.H. Unger, "Minimizing the Number of States in Sequential Switching Functions", IRE Trans. on Electronic Computers, Vol. EC-8, 1959, pp. 356-367.

[Pineda89] Jose Pineda de Gyvez, J.A.G. Jess, "On the Definition of Critical Areas for IC Photolithographic Spot Defects", Proceedings 1st European Test Conference, 1989, pp. 152-158.

[Pineda91] Jose Pineda de Gyvez, "IC Defect-Sensitivity, Theory and Computational Models for Yield Prediction", PhD report Technical University of Eindhoven, 1991.

[Rao89] V.B. Rao, D.V. Overhauser, T.N. Trick, and I.N. Majj, Kluwer Academic Publishers, Norwell, Mass., 1989.

[Riessen89] R.P. van Riessen, H.G. Kerkhoff, A. Kloppenburg, "Design and Implementation of a Hierarchical Testable Architecture using the Boundary Scan Standard", Proceedings 1st European Test Conference, 1989, pp. 112-118.

[Rijk91] Rolf Rijk, Rob Dekker, Hans Kerkhoff, "Self-Test of a 256kx4 bit stand-alone Static RAM", Proceedings 2nd European Test Conference, 1991, pp. 11-16.

[Rodriguez91] R. Rodriguez-Montanes, J.A. Segura, V.H. Champac, J. Figueras, J.A. Rubio, "Current vs. Logic Testing of Gate Oxide Short, Floating Gate and Bridging Failures in CMOS", Proceedings IEEE International Test Conference, 1991, pp. 510-519.

[Roth89] W. Roth, M. Johansson, W. Glunz "The BED Concept - A Method and a Language for Modular Test Generation", VLSI 89, Munich, 1989, pp. 143-152.

[Sachdev93] Manoj Sachdev and Math Verstraelen, "Development of a Fault Model and Test Algorithms for Embedded DRAMs", Proceedings IEEE International Test Conference, 1993, pp.815-824.

[Sas93] Jos van Sas, Francky Catthoor, Hugo J. de Man, "Test Algorithms for Double-Buffered Random Access and Pointer-Addressed Memories", IEEE Design & Test of Computers, June 1993, pp. 34-44.

[Shen83] J.P. Shen and J. Ferguson, "Easily-Testable Array Multipliers", Proc. of the IEEE Fault Tolerant Computing Symposium, 1983, pp. 37-40.

[Shen85] J. Shen, W. Maly, F. Ferguson ,"Inductive Fault Analysis of CMOS Integrated Circuits", IEEE Design & Test of Computers, December 1985, pp. 13-26.

[Smith80] J.E. Smith, "Measures of the Effectiveness of Fault Signature Analysis", IEEE Transactions on Computers, Vol. C-29, no.6, June 1980, pp. 510-514.

[Soden89] Jerry M. Soden, Charles F. Hawkins, "Electrical Properties and Detection Methods for CMOS IC Defects", Proceedings 1st European Test Conference, 1989, pp. 159-167.

[Somenzi85] Fabio Somenzi, Silvano Gai, Marco Mezzalama, Paolo Prinetto, "Testing Strategy and Technique for Macro-Based Circuits" , IEEE Transactions on Computers, Vol. C-34, No. 1, January 1985, pp. 85-89.

[Stanat77] Donald F. Stanat, David F. McAllister, "Discrete Mathematics in Computer Science", Prentice-Hall International Editions, 1977.

[Stans89] Rudi Stans, "The Testability of a Modified Booth Multiplier", Proceedings of the 1st European Test Conference, 1989, pp. 286-293

[Suk81] D. Suk and S. Reddy, "A March Test for Functional Faults in Semiconductor Random Access Memories", IEEE Transactions on Computers, Vol. C-30, no. 12, December 1981, pp. 982-985.

[Swaving88] Sieger Swaving, Alfred Ketting and Albert Trip, "MOS-IC Process and Characterisation within Philips", IEEE Proceedings on Microelectronic test Structures, Vol. 1, No. 1, February 1988, pp. 180-184.

[Syrzycki87] M. Syrzycki, "Modelling of Spot Defects in MOS Transistors", Proceedings of the IEEE International Test Conference 1987, Paper 6.1, pp. 148-157.

[Thatte77] S. Thatte, J. Abraham ,"Testing of Semiconductor Random Access Memories", Proceedings of the 7th Annual International Conference on Fault-Tolerant Computing, LA, USA, June 1977, pp. 81-87.

[Thijssen89] A.P. Thijssen, H.A. Vink, C.H. Eversdijk, "Digital Techniques", Edward Arnold, London, 1989.

[Thijssen90] A.P. Thijssen, H.A. Vink, C.H. Eversdijk, "Digitale Techniek, deel 2", Delftse Uitgevers Maatschappij, 1990.

[Thijssen93] Loek Thijssen, Frank Bouwman, Hans Vink, "Redundant States in Test Control Block Design", Proc. 3rd European Test Conference, Rotterdam 1993.

[Tou74] J. Tou and R. Gonzalez, "Pattern Recognition Principles", Addison-Wesley, pp. 90-104, 1974.

[Trischler80] Erwin Trischler, "Incomplete Scan Path with an Automatic Test Generation Methodology", Proceedings IEEE International Test Conference, 1980, pp. 153-162.

[Veenstra88] P.K. Veenstra, F.P.M. Beenker, J.J.M. Koomen, "Testing of Random Access Memories, Theory and Practice", IEE Proc. vol 135, pt. G. no 1,. February 1988, pp. 24-28.

1,. February 1988, pp. 24-28.

[Verhelst89] Bas Verhelst, "The use of a Test Specification Format in Automatic Test Program Generation", Proceedings of the European Test Conference, Paris, 1989, pp. 362-368.

[Walker92] D.M.H. Walker, "Tolerance of Delay Faults", Research Report no. CMUCAD-02-45, Carnegie Mellon University, Dept. of Electrical and Computer Engineering, August 1992.

[Wang86] L.T. Wang and E.J. McCluskey, "Complete Feedback Shift Register Design for Built-in Self-Test", Center for Reliable Computing, Technical Report No.86-17, 1986, Stanford University, CA 94305, USA.

[Weste93] N.H.E. Weste, and K.Eshrachian, "Principles of CMOS VLSI Design", Addison-Wesley, Reading, Mass. 1993.

[Williams73] Michael J.Y. Williams, James B. Angell, "Enhancing Testability of Large-Scale Integrated Circuits via Test Points and Additional Logic", IEEE Trans. on Computers, Vol. C-22, No.1, January 1973, pp. 46-60.

[Williams83] Thomas W. Williams, Kenneth P. Parker, "Design for Testability - A Survey", Proceedings of the IEEE, Vol. 71, No. 1, January 1983, pp. 98-112.

[Woudsma90] R. Woudsma, F. Beenker, J. van Meerbergen, C. Niessen, "Piramid: An Architecture-driven Silicon Compiler for Complex DSP applications", Proceedings IEEE ISCAS Conf., May 1990.

[Zegers90] J. Zegers, P. Six, J. Rabaey, H. de Man, "DGE: Automatic Generation of Controllers in the Cathedral-II Silicon Compiler", Proceedings of the European Design Automation Conference, 1990, pp. 617-621.

Index

Address decoder 44,51
Address generator 68,73
Application protocol 26
Application-mode testing 5
Architectural synthesis 22
Bit error number 59
Boundary Scan 82,112
Built-In Self-Test 64
Clock driver 189
Clock skew 180
Clock skew margin 183,187
Clock tree 191
Cluster 60
Cover property 157,158
Coverage
 defect coverage 20
 fault coverage 20
Critical area 14,50
Critical path lenght 50
Data background 57
Data generator 68,71
Data lock-out timing 189
Data transfer 180
Defect
 catastrophic defect 12
 global defect 11
 global defect 11,44
 non-catatrophic defect 12
 spot defect 11,44
Defect density distribution 12
Defect monitor 11
Defect size distribution 12
Defect-oriented testing 11
Defined logic signal 172
Delay model
 dynamic delay model 175
 static delay model 175
Delay time
 maximum propagation delay time 174
 minimum propagation delay time 174
Design centering 10

Design entity 151
Design hierarchy 30
Design partitioning 23
Design-for-test 2
Distributed test control 125
Evaluation plan 34
False path 175
Fault
- data retention memory cell fault 46,49
- fault 12
- fault model 20
- fault model 5
- fault propagation 53
- memory cell fault 46
- multiple access memory cell fault 46,49
- state coupled memory cell fault 46,49
- stuck open memory cell fault 46,68
- stuck-at memory cell fault 46,48
- transition memory cell fault 46,48

Fault simulation 21
Flip-flop
- negative pulse-triggered flip-flop 178
- positive edge-triggered flip-flop 177,179

Fully specified finite state machine 119
Functional control path 107
Functional data path 107
Functional testing 5
Hold interval 177
I_{DDQ} monitoring 14,61,78,79
Inductive Fault Analysis 13,48
Input assignment 119,132
Input distance 90
Inspection 11
K-factor 157
Leaf-macro
- leaf-macro access 81
- leaf-macro access paralellism 139
- leaf-macro access path 81,107
- leaf-macro input access path 141
- leaf-macro output access path 141
- leaf-macro state diagram 114
- Leaf-macro 24,76

Logic level 172

Macro
macro ... 24
virtual macro ... 31
Macro Test ... 6,19,24
March
13n march algorithm ... 58
6n march algorithm ... 58
9n SRAM march test algorithm ... 55
march element ... 54
march test algorithm ... 54
march test algorithm length ... 54
Maximum compatible subgraph ... 163
Mealy machine ... 118
Memory
memory cell array ... 44,46
memory test algorithm ... 42
Minimal clock period ... 186
Minimum-maximum gate level timing model ... 172
Monitor wafer ... 12
Moore machine ... 118,119
Multiple input signature register ... 67,75
Noise margin ... 173
NP-complete ... 92
Output distance ... 90
Parameter testing ... 6
Pattern applying register ... 85
Pattern observing register ... 85
Placement
deterministic placement ... 14
randomized placement ... 14
Process control module ... 11
Process Monitoring Testing ... 15
Pulse-triggered timing ... 178
Random access memory ... 42
Read/write logic ... 44,52,53
Reference level ... 173
Reset requirement ... 119
Resource
compatible resources ... 156
control resource ... 152
generated test plan resource ... 152
generated test plan resource set ... 153
input resource ... 152

output resource 152
resource 152
resource allocation 140
Response analyzer 68,74
Scan
full scan 82
partial scan 82
scan cell 83
scan chain 83
scan design 83
scan register 83
Scan chain
scan chain configuration 87
scan chain configuration cost 92
scan chain cost 91
scan chain routing 81.164
scan chain routing cost 88
scan chain routing problem 86,87
scan chain routing problem complexity 92
Scan register
farthest input scan register 97
farthest output scan register 97
scan register properties 97
scan register weight 99
Scan register length 90
Scan test application protocol 85
Scope
TCB scope 136
test control signal scope 136
Self-test controller 68.69
Setup interval 117
Signature 16
Silicon compiler 21
SRAM test algorithm 54
State
compatible states 123
equivalent states 122
state assignment 124,134
state merging 124,132
successor state 122
unused state 128
Stirling Numbers 94
Structural testing 4

TCB
TCB finite state machine 116
TCB FSM structure 116
TCB reset part 116
TCB reset state 116
TCB selection code 126
TCB selection part 116
TCB test plan code 126
TCB test plan part 116
Technology centering 10
Test algorithm score 62,64
Test assembly
sequential test assembly 142
test assembly 20,139,164
Test compatibility graph 162
Test Control Block
Test Control Block assembly 115,131
Test Control Block generation 34,164
Test Control Block 34,108
Test control path 107
Test control signals 81
Test control words 114
Test data path 107
Test objective 2,76
Test pattern 5
Test pattern generation 21,77,165
Test pattern insertion 29
Test plan
conditional test plan 28,32
generated test plan 27
initial test plan 26
test plan compatibility 159
test plan execution 140
test plan mapping 114,131
test plan paralellism 140,145,154,159
test plan step 140
test plan generation 27,164
test plan step 140
test plan step paralellism 141,146,155,160
test plan step compatibility 160
Test protocol 25
Test specification
device test specification 20

leaf-macro test specification 29
test specification step 140
test specification step paralellism 141,149,155,162
test specification step compatibility 161
Test strategy 26
Test synthesis 30
Testability 6
Testability synthesis 30
Threshold interval 172
Time disjoint 157
Time function 155
Timing margin 177
Transfer information 32
Transition time 172
Transmission delay 171
Two phase clock system 188
Undefined logic signal 172
Value function 155
Verification testing 3
VLSI MIX monitor 12
Wait counter 68,73
Wiring delay 171
Zero defects 2
Zero escapes 2

MIX
Papier aus verantwortungsvollen Quellen
Paper from responsible sources
FSC® C105338

If you have any concerns about our products,
you can contact us on
ProductSafety@springernature.com

In case Publisher is established outside the EU,
the EU authorized representative is:
Springer Nature Customer Service Center GmbH
Europaplatz 3, 69115 Heidelberg, Germany

Printed by Libri Plureos GmbH
in Hamburg, Germany